Routledge Revivals

A History of Birds

First published in 1954, *A History of Birds* by James Fisher is a pioneering exploration of ornithology, tracing the evolution of bird studies from manuscript traditions to modern scientific approaches. Covering topics like systematics, geographical distribution, speciation, bird populations, and their relationship with humans, Fisher provides a comprehensive analysis of avian history and ecology. Enhanced by detailed observations and a bibliography, this book remains an essential resource for bird enthusiasts and scholars of natural history.

A History of Birds

James Fisher

Routledge
Taylor & Francis Group

First published in 1954
by Hutchinson University Library

This edition first published in 2026 by Routledge
4 Park Square, Milton Park, Abingdon, Oxon, OX14 4RN

and by Routledge
605 Third Avenue, New York, NY 10158

Routledge is an imprint of the Taylor & Francis Group, an informa business

A Library of Congress record exists under LCCN: 54013164

ISBN: 978-1-041-35293-8 (hbk)
ISBN: 978-1-003-79702-9 (ebk)
ISBN: 978-1-041-35294-5 (pbk)

Book DOI 10.4324/9781003797029

Birds as Animals

I.

A HISTORY OF BIRDS

by

JAMES FISHER

M.A., F.L.S., M.B.O.U., C.F.A.O.U.

HUTCHINSON'S UNIVERSITY LIBRARY

Hutchinson House, London, W.1.

Melbourne Sydney Auckland Bombay

Cape Town New York Toronto

Birds as Animals, published by Wm. Heinemann in 1939, was the foundation for this completely revised and rewritten book, first published 1954.

Printed in Great Britain by
William Brendon & Son, Ltd.
The Mayflower Press (late of Plymouth)
at Bushey Mill Lane
Watford, Herts.

In memory of
MY FATHER

CONTENTS

PREFACE

THE return of Ornithology to the respectable sciences must have a lot to do with the rise of animal ecology, and the consequent recovery of an interest in field natural history at universities. The recent uniting of the Bureau of Animal Population and the Edward Grey Institute of Field Ornithology, in a Department of Zoological Field Studies under the Linacre Professor of Zoology at the University of Oxford; and the remarkable development of bird behaviour studies at the Cambridge University field station at Madingley, are all symptomatic of modern trends. Across the Atlantic fine new studies in anatomy and systematics, embracing knowledge obtained in the field, march with great advances in migration study on this side of the ocean. Every generalized book such as this is out-of-date the moment it is written.

This book is designed (as requested by the Biological Editor) for university students. As a consequence I have, after some hesitation, given orthodox citations of the many original research papers to which I refer; though I believe the time has come when students should be trained to do their own rummaging in the literature, as they are in the discipline of History. A great illusion of scholarship can be created by a ponderous bibliography, and no such thing is intended in this book.

In further apologia, I must explain the history of this volume. It rises from *Birds as Animals*, published in 1939. I soon found, when I tried to make a new edition of this book, that a new book entirely could be the only thing: so a new book this is, and, in scope, one covering only half the subjects of its predecessor. Those aspects of ornithology that can be arranged under the headings "migration and habits" must come in a second volume: in this there is space only to give an account of the observers and the observed, the bare bones of our science of ornithology.

It is often said by present-day ornithological pundits that

"the gulf between amateur and professional ornithologists has now disappeared". If readers of what I have to say about the history of birds and their watchers agree with me that there never was much of a gulf, I shall not have written in vain. The historical approach to ornithology, so beautifully exemplified by E. Stresemann's *Entwicklung der Ornithologie* (1951), reminds us always that poets and prelates, just as much as scientists and scholars, become fascinated and obsessed by birds, those odd and observable creatures of our external world, and never rest until they can find the truth about them.

Besides the kind friends and relations whose help I acknowledged in the 1939 edition of this book, I have also to thank Messrs. Witherby's, proprietors of *British Birds*, for permission to reproduce Fig. 2 (p. 147); and Professor H. Munro Fox, Biological Editor of the series, for his encouragement.

JAMES FISHER,
ASHTON
(M.S., April 1953)

CHAPTER I

BIRD HISTORIANS— THE AGE OF MANUSCRIPT

THE earliest-known ornithological records belong to the Old Stone Age—to the oldest civilization of the Upper Paleolithic. The Aurignacians, as visitors to the scene of fantastic beauty recently unearthed in the Lascaux Cave, near Montignac, in France, may see, were artists of the highest skill. Upon the walls of caves they engraved, inscribed or depicted the *Proceedings of the Aurignacian Magic Circle*, or the *Minutes of the Perigordian Hunting Society*; and among their 'documents', some of which were certainly holy, and others probably educational and scientific, are a few birds.

The civilization of Aurignac occupied south and central Europe from the line Wales-Kiev to Altamira beyond the Pyrenees, from about 16,000 to 11,000 B.C. One Pyrenean cave, Gargas, whose walls are patterned with the outlines of over 200 human hands (many of them oddly mutilated) has a beautiful crane or heron, which may have been figured in the early period of the Aurignacian civilization, 17,000 or 18,000 years ago, and is thus probably the oldest bird record. Other Aurignacian birds have been found on the walls of a cave at La Vilera de-El Arab (a stork) and among the great battle frieze of Minateda, near Albacete, in Spain.

The bird-topped stave in the famous 'prehistoric tragedy' on the wall of the crypt at Lascaux is part of a picture painted in the later Magdalenian occupation of the cave. Very few birds are known from the work of late Upper Paleolithic men; the Solutreans, who succeeded the Aurignacians, were carvers and sculptors, not painters; and nobody has yet found a bird-carving of theirs. The wonderful Magdalenians, the last of the Old Stone Age, who flourished from about 9000 to 6000 B.C., excelled in both graphic and plastic arts; their birds are rare, and their finest known examples are a lively spear-thrower carved from reindeer-antler in the form of a capercaillie, a gosling engraved on reindeer-antler, and a swan engraved on a pebble. At El Pendo in northern Spain a Magdalenian rock-engraving represents what was almost certainly intended to be two great auks.

About 8000 years B.C. the amelioration of the climate after the last great glaciation became marked. The Baltic became a sea, no longer permanent ice. The Magdalenians went north, with the reindeer they were, as hunters, associated with, and probably became the ancestors of the northern hunting tribes, notably the Eskimos; in their place came the agricultural men of the New Stone Age with their polished flints. To the early Neolithic culture belonged the inhabitants of the Tajo Segura, a cave in southern Spain on whose walls are the outlines of ten or twelve recognizable genera of birds; great bustard, crane, duck, goose, raven, spoonbill, flamingo, purple gallinule, glossy ibis, white stork, and probably eagle and harrier, all of which are found in Spain today (Verner, 1914). These figures have been dated between 6000 and 4000 B.C.

The world's first civilization (already centuries old by 3500 B.C.) was that of the Sumerians of Mesopotamia; they were architects and artists while Egypt was still divided into barbarous tribes. The oldest known Sumerian edifice in Mesopotamia was a 'gay and fanciful' building, as Woolley (1942) describes it, at al 'Ubaid, about four miles from Ur. This building, which dates from about 3100 B.C., was decorated, amongst other things, by a mosaic frieze of stone and mother-of-pearl, which contains a row of birds, probably doves, roughly cut in white limestone. Also in the frieze was a lion-headed eagle attacking a bull (the illustration of some legend). And upon the building was also a great copper relief representing a near-life-size eagle grasping two stags. Another object of this date, the First Dynasty of Ur, is an alabaster relief of a boat with a goose in it.

The earliest Egyptian bird painting is not much younger; upon a fresco of about 3000 B.C. on the tomb of Ne-fer-Maat at Medum are depicted six geese of three species. One pair of these is either *Anser erythropus*, the lesser whitefront, or

A. albifrons, the whitefront; probably the latter, for although the white fronts of the Medum birds extend rather far back on their foreheads, their big beaks and somewhat gross build clearly suggest the larger species, which is at present the commoner of the two as a winterer in Egypt. Another pair are *Branta ruficollis*, the red-breasted goose, which has only once been recorded from Egypt in recent years, but which may well have had a different distribution 5000 years ago. The other two geese are bean-geese, *A. arvensis*; there is one modern record of this species in Egypt.

During the three millennia B.C. avian decorations on the buildings and ornaments of Egypt and Assyria grew profuse; many species appear in hieroglyphs, and many were mummified as sacred animals. R. E. Moreau (1930) recognizes no less than ninety species of Ancient Egypt, and shows that the fauna of Egypt in those years was considerably different from what it is at present.

Birds are common in the art of the ancient Greeks, on coins, statuary, tombs, palaces, sceptres and ceramics (Thompson, 1936), eagles, the lammergeier, vultures, the cock, cranes, owls (the owl is the badge of Athens), the hoopoe, herons (symbol of Athene), ibises, hawks, the wryneck, the cuckoo (on Hera's sceptre), the raven, the crow, the swan (of Apollo), guinea-fowl, doves, the quail, the stork, the peacock, the swallow.

In the writings of classical Greece Aristotle [384–322 B.C.], the pupil of Plato and tutor of Alexander of Macedonia, stands out as perhaps the first and last person who could write down the sum of human knowledge, including everything that anybody of his time 'knew' about birds. In those of his works that survive he mentioned about 170 species, of which—depending upon the standards of the interpreter—perhaps not more than 126 are recognizable. He said a good deal about their physiology and reproduction, and even their ecology and, as Nicholson (1929) claims, their territory.

The best of the ornithologists of the Ancient Graeco-Roman civilization appears, however, not to have been Aristotle, but Alexander the Myndian (first century), some but not all of whose work appears to have been borrowed from the *Ornitho-*

logonia of a mysterious and little-known figure, Boios. Alexander the Myndian appears to have been a fine bird-watcher; he distinguished between the house- and the tree-sparrow, identified and pointed out a close relationship between the garden-warbler and blackcap and even went so far in scientific observation as to investigate the legend of the song of the dying swan, by practical field-work—he followed many, but never heard one sing! (Douglas, 1928).

Relatively few birds are mentioned in the Bible (Tristram, 1911); only twenty of those of the Old Testament can be run down to certain species, of which eight are birds of prey (including owls). The 'sparrow' is a general term to cover small birds; and there are eight categories like 'hawk' and 'heron', which each cover several species. Besides 'sparrow' the birds of the New Testament are the domestic fowl (*not* mentioned in the Old Testament), the griffon-vulture (*nesher* or *niser*, the 'eagle'), raven, turtle-dove and rock-dove (with its descendant the domestic pigeon).

Aristotle made a beginning of the classification of birds in a natural system forming eight groups. His most scientific successor, a Roman army officer and friend of Vespasian, Caius Plinius Secundus, Pliny the Elder [A.D. 23–79], divided them into three. That Pliny improved upon Aristotle seems clear to most who study the lucid comparison of their knowledge offered by William Turner in 1544 (A. H. Evans, 1903), though it did not appear so to Alfred Newton (1884). But Claudius Aelianus (Aelian) who flourished and wrote in about A.D. 140, contributed little new; indeed he ushered in 1,000 years of anecdotage—years of fables and bestiaries, of confusion of fable with fact, of opinion with fact, of religious doctrine with fact. To quote Casey Wood (1931, p. 3),

> "the early Christian writers were far removed from scientific methods, and instead of consulting Aristotle and other observant naturalists they quoted Aelian and similar works of the imagination in which scraps of folk-lore, travellers' tales, and fragments of misapprehended science were set forth in what was regarded as an elegant and authoritative style".

Some of these works of early writers, in spite of their scientific triviality, are worth mentioning as human documents, though others, like the *Collection of Marvels* of C. J. Solinus [*c.* A.D. 275], are simple compilations (this one mainly from Pliny). The *Physiologos*, which started as a written (and for hundreds of years copied, edited, added to and altered) compilation in Alexandria or Syria in about 370 is a masterpiece of confusion; mystical, magical and moral. It was as popular as it was preposterous: its chief contribution (admittedly a good one) to civilization is the gargoyle. Among the more amusing perpetrations of the *Physiologos* (as compiled by Wood) are the following: the plover presages the death or recovery of patients; the pelican recalls its young to life with its own blood; the eagle renews its youth by sunlight and bathing in a fountain; the hoopoe redeems its parents from the ills of old age; the partridge hatches eggs of other birds; the vulture is assisted at birth by a stone with a loose kernel; the swallow brings forth but once; the heron touches no dead body and keeps to one dwelling-place. It was the *Physiologos* that gave rise to the Dialogues, Fables and Bestiaries that were the civilized world's substitute for Natural History for 1,000 years; how wide the gap between truth and fable became can be determined by reference, amongst many others, to the *Etymologiae*, compiled in about 602 by Bishop (later Saint) Isidore of Seville [*c.* 560–636], or to the *Sermonum* of Rabanus Maurus, Archbishop of Mainz [776–856].

It was largely the Arabic-writing Muslim doctors that brought the world back, if not quite to reality, at least to Aristotle. Al-Jāhiz [800–865] and Aboru ben Masoweih (or Mesuë, d. *c.* 857, wrote in 840) each wrote a *Book of Animals*. These works, those of Aristotle, and an interest in the external world inspired the important studies of Abu Ali Husain ibn Sīnā (or Avicenna, 980–1037), Abu'l Welid Muhammad ibn Rushd al-Maliki (or Averroës, 1126–98; wrote 1170) and Zakariÿa ben Muhammad Al-Qazwīnī who wrote the *Wonders of Creation* in about 1250. Nordenskiöld (1929) says that

"the Arabic commentators, Avicenna, Averröes and others, were the first to act as guides to an understanding of the treatises on nature (of the Ancient Civilization) and to help Europeans to penetrate that world of phenomena whose existence they had entirely forgotten".

It was by the thirteenth century that the impact of the Arabians and their modernized Aristotle began to be felt. It was in the time of Frederick II [1194–1250], Emperor of Germany, that new writers, notably Dominican friars, began carefully and subtly (for there was mortal danger in any overt flouting of the doctrines of the church) to introduce reality—or at least some elements of the world of nature—into their philosophical books. The most notable of these was Albert von Ballstadt, or Groot [1193–1280] who became a professor in Paris and later bishop of Regensburg. He is usually known as Albertus Magnus, and wrote his *De Animalibus* in about 1260–70. Newton (1884) is unduly scathing about this book for, as Wood points out, Albertus

> "kept alive and encouraged an interest in nature herself at a time when the minds of men were powerfully and exclusively directed to a supernatural world in or beyond the skies".

Other Dominicans, contemporary with Albertus, who wrote on nature were Vincent de Beauvais [d. 1264]; and Thomas de Cantimpré from Brabant [1186–1263], whose *De naturis rerum* was progressive. Vincent's *Speculum Quadruplex* was a voluminous work, of which the seventeenth volume was a dictionary of birds. Frederick himself, assisted by his son King Manfred of Italy [*c.* 1232–66], prepared a classic codex (in the Vatican library) on falconry and natural history, which was edited and amended by Albertus Magnus. It is beautifully decorated with portraits of many species of wildfowl and birds of prey (see Wood, 1942).

It was in the fourteenth century that the coming of the Renaissance found a few heralds. Among them was another Dominican, Konrad von Megenburg [1309–78] whose celebrated *Buch der Natur* was finished in about 1346; he clearly had doubts about the doctrines of the past.

It was much easier for travellers and poets to come to grips with reality than scholars—at least until the sixteenth century. This is nowhere more evident than in Britain: if we want to find out what British birds were known in those days we do not go only to the ecclesiastical or academic manuscripts, but also (and mainly) to poets, travellers, story-tellers, housekeepers, lawyers, hunters: to the noble unknown authors of such eighth-century wonders as *Beowulf*, *The Perils of the Seafarer* and the *Anglo-Saxon Chronicle*; to the tenth-century *Colloquy* and *Vocabulary* of Archbishop Ælfric, and the laws of King Howel of Cambria; in the eleventh century to household accounts of the monks of Rochester and Waltham, the *Book of Ely*, and great *Domesday Book* itself; in the twelfth to the travels of the indefatigable Welshman Gerald of Barry, called Giraldus Cambrensis.

Many documents of the thirteenth century identify edible birds in rent-rolls and household accounts; the first record of the crossbill (1251) is as an orchard pest in St. Albans, and realistic birds appear in the lines of such fine poems as *The Owl and the Nightingale* (*c.* 1216–25), the famous *Cuckoo Song* of *c.* 1250 ("Sumer is icumen in, Lhude sing cuccu!") and *Spring-tide* (*c.* 1300: "The threstlecoc him threteth oo, Away is huere wynter wo"). In the fourteenth century we find many birds in price-lists and catalogues; but most realistically, and with a true understanding of their nature and life, in Geoffrey Chaucer's *Parlement of Briddis* and *Romaunt of the Rose*. In the fifteenth century the external world began to creep into the pages of ecclesiastical manuscripts, so that we have at last a harvest of nature notes from them as well as from the rest. Dame Juliana Berners [1388–1485] of Sopwell (Hertfordshire) wrote *The Book of St. Albans*, on hawking, hunting, fishing and armour, in about 1460; it has much on huntable birds. This century was one of bird poems (e.g. Sir Richard Holland's *The Howlet*), vocabularies (e.g. the *Nomina avium fferorum*), bird art (e.g. the wonderfully naturalistic drawings of Pisanello and Leonardo), travels (e.g. the visit of the Venetian ambassador Capello to London and that of William Botoner to Cornwall and Wales) and feasts of a gargantuan kind such as Henry IV's wedding breakfast (1403), Henry V's coronation dinner (1413), Henry VI's coronation dinner (1422), the inauguration dinner of Neville as Archbishop of York (1465) and that of Wareham as Archbishop of Canterbury a few years later, and the Duke of Buckingham's wedding breakfast (1480). At all these fantastic meals the tables groaned under the weight of birds; according to one account at least 14,520 birds were ordered for Neville's banquet, belonging to at least twenty species.

Largely assisted by the useful *Early Annals of Ornithology* of J. H. Gurney (1921) and various bibliographies, I have tried to find out how many birds have been described as present in Britain in the documents of the Age of Manuscript. This—as far as birds are concerned—continued until the end of the fifteenth century, though the first printed book on animals was published before 20th July, 1467 in Strasburg (an edition of the *Sermonum Proprietate* of Rabanus Maurus). By 1500 eighty species have been mentioned, though in some cases, e.g. *swan*, *eagle*, *owl*, *tern*, I have had to assume the particular species intended, and several remain quite indeterminable.

The 'first British Birds' are all appropriate characters of romantic-heroic-travel poems; the eleven I can find of the eighth century[1] are the raven (*Beowulf* and *Anglo-Saxon Chronicle*) eagle (*B.* and *Seafarer*, presumed to be the white-tailed eagle), swan (*B.* and *S.*, presumed to be the whooper), gannet, cuckoo, gull (*The Husband's Message*, a poem in the *Exeter Book*), tern (the common and arctic terns were not distinguished until 1819), whimbrel or curlew (it is just as likely, I believe, that the Hu-ilpe of the *Seafarer* refers to the northern bird as to the curlew), goshawk, kite and crane. In *Beowulf*, the sea is alluded to as 'ganotes bæð', which is a fine poetic conceit which certainly puts things in their right proportion! "Over the gannet's bath many a one shall greet the other with gifts".

The transition from the ten birds of the eighth to the eighty of the fifteenth century was a gradual one, as far as I can discover, and no writers, except possibly Ælfric and Chaucer, and certainly Giraldus Cambrensis, contributed more than a handful of species to the 'list'. Ælfric's *ficedula*, or

[1]St. Serf, who lived at Culross in the sixth century, had a tame robin, according to a beautiful story quoted by Lack (1950). But this story is not provedly based on an early document, and may have been first written in 1125.

swertling may be the blackcap; if so, it is the only British warbler mentioned before 1500. Other bird families whose members were not distinguished from each other (or of which only one member was recognized) in the Age of Manuscript were the finches, buntings, sparrows, larks, pipit-wagtails, tits, shrikes, swallows, woodpeckers, owls, cormorants, grebes, plovers, terns, gulls and rails; and I can find no record at all of flycatchers, tubenoses or divers.

CHAPTER II

BIRD HISTORIANS—
THE AGE OF PRINT: BRITISH BIRDS

IF example be needed of the impact of the Age of Print and Science upon the world, the transformation of the list of recorded British birds in the sixteenth century is a good one, for it was nearly doubled, rising, as I have it in my collection of names, from eighty to 138. More than half the fifty-eight 'new' birds were contributed by William Turner [1500–1568], the father of British ornithology, whose *Avium Praecipuarum* was published in Cologne in 1544, the first printed bird book; indeed, if we remove from his 130 recognizable birds those which are domestic or were not seen in Britain, we are left with 105 species (including two gulls and two grey geese whose final identification is not possible) of which no less than thirty-seven have appeared as British birds in no previous document, manuscript or printed, that I have seen or heard of. Turner recognized for the first time three finches and four buntings, three tits and several other passerines. His was the world's first printed book on birds.

Turner was a reformer, and in about 1540 was imprisoned for preaching without a licence; on his release he went to the Continent and became friends with and stimulated by Conrad Gesner of Zurich, and Gybertus Longolius of Utrecht, two other great naturalists of the scientific revolution. But when he was not in trouble with the established Church (and this changed a good deal in the times of Henry VIII, Mary and Elizabeth) Turner lived in England, where it is clear that he had much first-hand knowledge of birds, particularly marsh-birds. He knew that the hobby was a summer visitor, and described exactly its habitat and some of its principal prey; he had obviously watched closely the hen- and marsh-harriers, dipper, common sandpiper, black-headed gull, black tern, swallow, swift, martins, cormorants, great grey shrike, robin and redstart.

The next important name in the development of our knowledge of British birds is that of John Kay or Caius [1510–73], who was the second founder of a Cambridge college in 1557 and published his *De rariorum Animalium* in 1570. Kay's observations on birds were not many, but excitingly original; he kept a puffin in his house for eight months and made good observations on the food of the jackdaw. He was a colleague of Turner's and a correspondent of Gesner.

In the seventeenth century it seems that another sixty-four species became recognized as British, largely through the exertions of the fine naturalists Sir Thomas Browne [1605–82], Francis Willughby [1635–72] and John Ray [1627–1705]. Of these Ray was the greatest; indeed he is, in the opinion of most, the greatest field naturalist that has ever lived, and the founder of modern zoology: readers are referred to the admirable and definitive recent biography of him by C. E. Raven (1942) which was followed (1947) by a useful book on his predecessors.

Browne, who made the first British records of bearded tit, hoopoe, roller, Manx shearwater and little auk, amongst others, lived in Norwich from 1636 onwards. Most of his bird notes were given in correspondence with Christopher Merret [1614–95], a moderate naturalist whose *Pinax rerum naturalium Britannicarum* was first published in 1666; but many were sent to Willughby and Ray, and some not published until 1835. In 1660 Willughby and Ray started their first journey together, to north England and the Isle of Man, visiting Browne at Norwich later in the year. From then until Willughby died in 1672 Ray hardly missed a season travelling with his great friend, or with his pupil Skippon: they went into most parts of England and Wales, and many of Scotland, and to the Low Countries, France, Germany, Austria, Italy, Sicily and Switzerland. Ray found about half the 'new' birds of the century—thirty species or more; it seems likely from the Willughby-Ray *Ornithologia* (1676; English version 1678) that he was aware of the difference between the chiffchaff and wood-warbler—the identity of the third member of this difficult trio, the willow-warbler, was suspected by Derham (1718) and confirmed in 1768 by White (1788). Ray could distinguish redpoll and twite, yellow and grey wagtails, coal- and marsh-tits; and among his identifications in difficult families are four warblers, the little and probably the Sandwich terns, and the great blackback.

By the end of the century, thanks mainly to Browne, Merret and Ray, but also to the Scots Sibbald (1684) and Martin (1698) and a few others, the list of recognized British birds had been brought to 202. Martin Martin, from his visit to St. Kilda in 1697, brought the first detailed and accurate accounts of the lives of many sea-birds, of which the fulmar and the storm-petrel had never been previously recorded as British birds (Leach's petrel, which also nests on St. Kilda, was not distinguished from the storm-petrel until 1818).

In the next century, in spite of the spread of ornithology and natural history, only a further thirty-nine British species were added, mostly by William Derham [1657–1735], Eleazar Albin [*c.* 1713–59], George Edwards [1694–1773], Thomas Pennant [1726–98], Marmaduke Tunstall [1743–90], John Latham [1740–1837], Gilbert White [1720–93] and George Montagu [1751–1815]. This is because comparatively few distinctions of established but hitherto unrecognized species remained to be made, but there were not yet enough bird-watchers to encounter many rarities and vagrants. Distinctions *in Britain* that were cleared up in the eighteenth century included tree-sparrow, black- and red-throated divers, arctic skua, willow-warbler, sedge-warbler, Dartford warbler, lesser whitethroat, tree- and rock-pipits, greenshank, grey phalarope, lesser blackback, Slavonian grebe, wood- and purple sandpipers and rough-legged buzzard. From the point of view of general systematic ornithology, Latham was perhaps the most important eighteenth-century British writer, though he did not use the great new system of nomenclature of Linnaeus until 1790. From the point of view of British systematic ornithology, Thomas Pennant was perhaps the most important; his *British Zoology* first appeared in 1766 and went into edition after edition. But undoubtedly the dominant figure of the century has proved to be the modest curate of Faringdon in

Hampshire with whom Pennant corresponded at such length, Gilbert White of Selborne; the man who started us all bird-watching.

I have attempted elsewhere (White, 1947) to analyse the reasons for the success of the *Natural History of Selborne*, first published in December 1788, and of which, since then, there have been about 150 editions. Unquestionably White's personality and his marvellous literary skill, which gives an impression of simplicity but is the fruit of the endless pains of a genius, are among these reasons: but he excelled as a fieldman, and his original discoveries (leaf-warblers, lesser whitethroat, noctule bat, harvest-mouse) remain as interesting —as exciting, even—to us as they must have been to him. This humane, tolerant, kindly and modest man must have inquired into the nature of things with no other motive but that of inquiry or, as he probably thought of it, worship. There are many now who, in so far as they possess such motives, owe them to him. In White the nature-investigator and nature-lover were inextricably confused; and many believe that the pursuit of truth in natural history will continue to be successful only so long as this benign confusion is preserved.

The only other work with a claim on the hearts of bird-watchers comparable to that of the *Natural History of Selborne* is Thomas Bewick's great picture-book, *A History of British Birds*, which began to come out in Newcastle-upon-Tyne in 1797. This must be primarily judged as a work of art. The first volume, of 1797, the *Land Birds*, had a text largely compiled from the works of previous authors. It is of no importance when compared with Bewick's woodcuts, and was written, as a matter of fact, by Ralph Beilby to whom Bewick was apprenticed to learn engraving in 1767. The text of the *Water Birds*, which appeared in 1804, was not only illustrated by Bewick, but entirely written by him. The text is of no better quality than Beilby's text of the first volume, and proves nothing except that Bewick was not a *very* good observer of natural history or a very original thinker on the subject. But when he died in 1828, at the age of seventy-five, he had done for art what Burns did for poetry—put a little honest humanity in it.

By modern standards of bird illustration the figures of

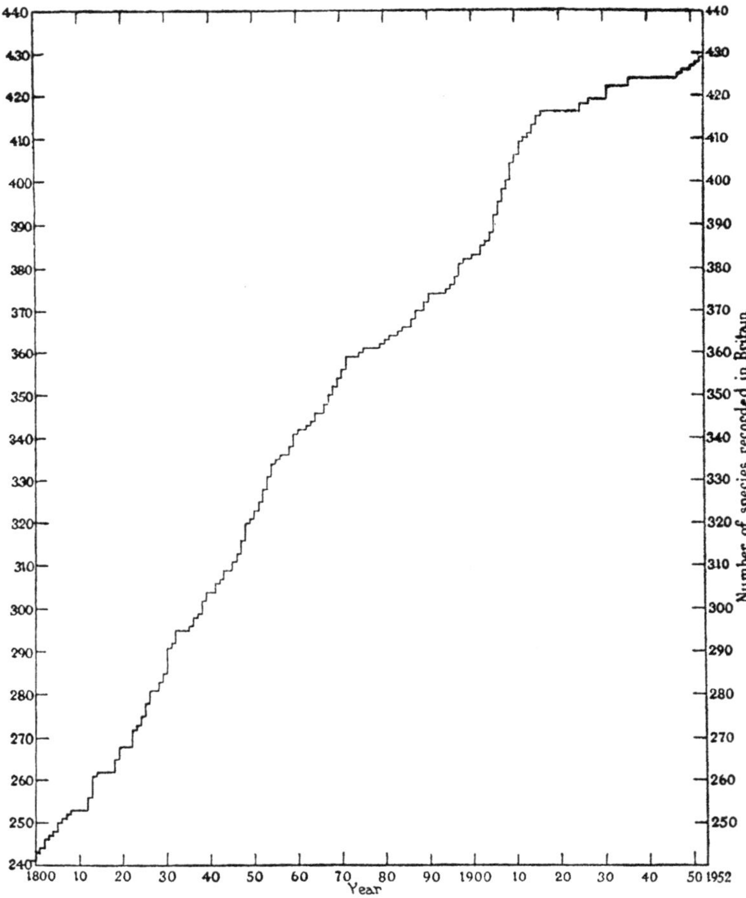

Fig. 1. *The increase, since* 1800, *of the total number of bird species recorded in the British Isles. The modern criterion of species has been used, and all old lists re-analysed. Records before* 1950 *as accepted by B.O.U.*

birds in some of Bewick's cuts are unnatural; with those he watched himself as a boy, he is at his best. His fieldfare, for instance, is surely based on a field sketch. His blackbird, with his old home at Cherry Burn by Tyneside in the background, may have been sketched at the very spot on which it appears in the cut. Bewick at his best was perhaps the best woodcut nature-artist since Dürer.

This digression on the subject of White and Bewick must be forgiven, for these heroes of bird-watching have had a special influence upon the development of ornithology, that has been felt beyond Britain. We must continue with our development of the list of known British birds, and our account of some of those who found them, for to follow the list is as good a way as any to understand British ornithology, and perhaps better than a succession of names and dates. From 1800 onwards I have established the year in which each new British bird was first reliably seen; the increase has continued, in a steep but not quite regular curve (Fig. 1) until the present day. From 241 species on the list in 1799 the total has risen to *c.* 429 in 1952. The curve has not behaved like a normal hyperbola, though it did so until about 1890, with the peak 'intake' of new species about 1830, and a subsequent decrease in each decade. I suspect that this was linked with the rise of and improvement in the shotgun. Towards the end of the nineteenth century, when shotgun collecting would have continued to give diminishing returns had the number of collectors not increased, the quick increase and development of ornithology as an amateur sport, and the improvement of books and sight-recognition, enabled twenty-three new species to be recorded for the decade 1901–10, as great a number as in any decade since 1799, and having been only equalled in 1821–30. Since this twentieth-century peak the returns have once more been diminishing, though no doubt they would have diminished more than they have had not bird-watchers continued to increase, indeed to multiply. Only three new species were recorded in the decade 1921–30, five in 1931–40, and three in 1941–50, and an extrapolation of the curve from here would suggest that only about eight new species altogether are likely to be observed in the second half of the present century, to compare with the forty-four in the first half. But in 1951-53 five or six new species were recorded, of which four were transatlantic drift-wanderers, and it is clear that the British List may have to be reassessed to embrace a rather large number of American passerines that have hitherto been rejected, in my opinion unreasonably.

The accessions to the British list in the early nineteenth century owe much to George Montagu, that energetic field-worker who in turn owed much to White's inspiration. ("Your work produced in me fresh ardour", he once wrote to White.) Montagu swept up, in his efficient way, almost the last of the birds that were unknown because unrecognized (usually undistinguished from some close relative), not unknown because rare or vagrant. In 1800 he distinguished the cirl-bunting from the yellowhammer for the first time in Britain, in 1802 the harrier that now bears his name. In 1813, in the *Supplement* to his famous *Ornithological Dictionary* of 1802 he gave the first description ever, lay or scientific, of the roseate tern, and recorded for Britain for the first time four other species.

A few species that really belonged to our regular avifauna remained unrecognized until later. Leach's petrel was first scientifically described in 1817 and discovered in Britain (at St. Kilda) by William Bullock in 1818. Next year, 1819, the fundamental difference between the arctic and common tern was at last scientifically cleared up by J. F. Naumann[1]. In 1822 Laurence Edmonston of Shetland identified the glaucous gull as a winter visitor to these islands, and in the following year found that the very similar 'Iceland' gull was also a British bird. In the next year, 1824, William Yarrell identified Bewick's swan as a British bird, it having previously been confused with the whooper. The marsh-warbler was not distinguished from the reed-warbler in Britain until 1871, when Edward Blyth discovered it near London—there is highly suggestive evidence, however, that it nested in Hampstead in 1861–63 and 1865 (Harting, 1866; Fitter, 1949). The last of the 'present but hitherto unrecognized' birds was the willow-tit—also found in Hampstead. In 1897 Kleinschmidt and Hartert were

[1]A rather similar case is that of the nightingale, which was not separated from the thrush-nightingale until 1831. However the latter has only twice been recorded in Britain, which has not therefore been affected by the confusion.

looking through a tray of skins labelled 'marsh-tit' in the Bird Room of the British Museum, when they recognized two from Hampstead to be undoubtedly willow-tits; and in the same year the Tring Museum obtained two willow-tits from Finchley. After that it took some years before it was finally discovered that the willow-tit had an even wider distribution in Britain than the other very similar species.

Ornithology has reached such a state of organization in Britain now that the identifications of three recent species—the isabelline shrike seen by Miss W. H. Flower and Miss I. M. Kinnear in 1950, the blue-cheeked bee-eater seen by Miss Hilda Quick in 1951 and the collared turtle-dove seen by Reg May and the writer in 1952 have all been accepted by editors of journals on sight records. Not long ago the birds would have been either shot or not identified.

* * * * * *

Note: There was no idea of an official British List until 15th May, 1878, when the British Ornithologists' Union agreed to make one at their Annual General Meeting. The work was entrusted to P. L. Sclater and N. T. Wharton with a committee of six others. From internal evidence it seems that they considered no records after 1880, though the list was not published until 1883.

It is interesting to compare it with what we now know of the birds listed, their records and records of other species to 1880. The committee listed 376 species, whereas the true number accepted by 1880 appears now to have been 363 (by modern critical standards). The discrepancy is accounted for in this way: they listed sixteen forms as full species which have since been united with other recorded British species (as subspecies), and nine species which have been subsequently dropped as invalid records (though some were readmitted on valid records after 1880); and they rejected ten species whose pre-1881 records have subsequently been accepted, and overlooked two species, the sociable plover and Siberian pectoral sandpiper, observed before 1880, for the very good reason that the record of the sociable plover, made in 1860, was not made known until 1888, while the first Siberian pectoral sandpiper, shot in 1848, was confused with the American pectoral sandpiper until the specimen was re-examined in 1892.

Many subsequent amendments, corrections and revisions of the British List have been made. The latest, published at the end of 1952, lists 426 species as acceptably recorded in Britain to 31st July, 1950 (Glegg and others, 1952).

CHAPTER III

BIRD HISTORIANS—THE AGE OF PRINT: WORLD ORNITHOLOGY

WE have used as an example of the history of ornithology since 1500 that country where bird-watching and ornithology are most intensively followed, and we have taken the history of Britain's list of birds to show how the observers and the observed have drawn closer to each other, so that eventually all indigenous species, or regular visitors, became 'known' or 'well worked-out' (to use two expressions which nowadays echo wherever ornithologists gather together) and the strayest of stray visitors stands a good chance of becoming recorded—and on an acceptable sight-record. But the successful identification and geographical plotting of birds is but one facet of ornithology, and we must now retrace our steps once more to 1500, to consider wider aspects of it in the wider world.

If William Turner was the father of modern ornithology, it is certain that his European naturalist friends and colleagues of the sixteenth century were its uncles, the Old Guard of the Revolution in Natural History. The Age of Print had been ushered in by some reminders of the Middle Ages such as the (*H*)*ortus Sanitatis* of Johann von Caub, a Frankfurt physician, which, printed in 1485 (the earliest German herbal), was a compilation of the usual kind from many sources, notably Isidore of Seville [570–636], Vincent of Beauvais [*c.* 1190–1264] and Konrad of Megenburg [1309–78], not to mention the *Physiologos* (first version *c.* 370). It was illustrated by some of the earliest woodcuts, and contained very little real natural history. The sixteenth century's contrast can be made vivid by comparing the woodcuts of Dürer with those of the *Ortus*, and the text of any of the works of Gilles, Turner, Gesner, Belon, Coiter, Aldrovandi or even Longolius and Wotton (though these two last are more derivative than the others) with that of von Caub; it was a contrast as remarkable as it was sudden.

In the early sixteenth century Leonardo [1452–1519] was at the height of his powers. Albrecht Dürer [1471–1528] was drawing his wonderful plants and animals—many must know the reproductions of his brown owl of 1508. Some of Dürer's drawings were contributed to the works of Conrad Gesner of Zürich [1516–65] who has rightly been described as the father of bibliography, and who among many other excellent books published his *Icones Avium* in 1555. Gesner was a great correspondent, and friends flocked to his house from all over Europe, among them Turner, Gilles and Belon; Pierre Belon of Le Mans [1517–64] also published a bird book, *L'Histoire de la Nature des Oyseaux*, in 1555; his was less scholarly but more original than Gesner's and had a more definite and intelligent system of classification. Belon and Volcher Coiter [1534–90] were the first real bird anatomists; Coiter's Nüremberg treatises (1573, 1575) provided one of the earliest attempts at a classification of birds based on internal characters, and contain exquisite studies of bird skeletons. Pierre Gilles, or Gilius, had published a work in 1533 based on medieval manuscripts but containing the first printed description of the elephant from nature; he is best known as the teacher friend of Belon (and, with Belon, tutor to Prince Henry, heir to Francis I of France), who accompanied him on a tour of the Aegean, Turkey, Asia Minor and the Levant in 1547–49 which resulted in a travel and natural history book (Belon, 1553), far in advance of its time in objectivity. In 1599–1603 the ornithological works of Ulisse Aldrovandi [1522–1605] of Bologna were published; but these compilations only slightly improved (if at all) on the books of Gesner and Belon; and this somewhat overrated scholar was overtly jealous of Gesner, which improved his objectivity no more than it increased the relevance of some of his passages.

From the beginning of the seventeenth century onwards it is necessary to divide ornithology into its constituent subjects if we are to understand its processes and progress. Some of these we must develop in subsequent chapters. Here let us start with Comparative Anatomy. It is the custom today amongst some to regard the work of the comparative bird-anatomists as of secondary importance. This is a great mistake, since, owing to the particular circumstances caused by the evolution of birds, and the fact that their classification is probably more difficult than that of any other group of animals, the real evidence which has decided the present classification has come almost entirely from the tireless labours of such people. Pliny classified birds on their feet; in those days it was probably as good a way as any. Willughby and Ray classified birds on their beaks as well as their feet, a method copied for many years after by the great reformer of natural history Carl Linné [1707–78], and even by the skilful and careful Baron G. L. C. F. D. Cuvier [1769–1832] and F. Tiedemann [1781–1861], whose *Tableau élémentaire de l'Histoire naturelle des Animaux* and *Anatomie und Naturgeschichte der Vögel* were published in 1798 and 1810–14 respectively. Indeed, it was only in the early nineteenth century that real advances began to be made in the comparative anatomy of what is—from the anatomical point of view—the most 'difficult' class of vertebrates. The eighteenth century did perhaps no more than shake off the idea, which persisted until 1681 (Gerard Blasius [1623–82]), that bats could be classed with birds on anatomical grounds. But in the nineteenth century took place most of the fundamental work upon which our present knowledge of the family tree of birds is based.

In many ways the most revolutionary step was taken by Blasius Merrem [1761–1824], who, in his later works (1816, 1819), separated the birds into two great divisions, carinate (keel-breasted) and ratite (flat-breasted). But his diagnosis of the difference between these groups of birds did not rest on this characteristic alone. He found eight other characters of first-class importance, such as the presence or absence of hooks on the barbs of feathers, the position of various wing-bones, the number of vertebrae, and so on.

The earlier work of C. L. Nitzsch [1782–1837] was osteological, and some of it (1811) remarkable, in that it separated two birds, the nightjar and swift, from their apparent relations, the swallows and martins, on points of pure anatomy. But though Nitzsch also made pioneer investigations of the blood-vessel system of birds, his chief contribution to comparative anatomy was through his entirely original studies of feather-tracts; his *Pterylography* only began to be published in 1833 and did not fully reach the world until 1840: and it is true to say that his important researches were scarcely recognized and honoured outside his own country, Germany, until the end of the nineteenth century.

Another important anatomist was Etienne Geoffroy Sainte-Hilaire [1772–1844] whose several works, appearing in the first third of the nineteenth century, concerned particularly the ossification of the avian skull (1807, 1824), breast-bone and vocal organs, and the intestinal canal and reproductive organs. Perhaps his most valuable work (1833) was his tracing of the breast-bone from five centres of ossification. Another important anatomist, also interested in the sternum of birds, was F. J. l'Herminier, a Creole of Guadeloupe in the West Indies who worked in the Paris Museum and published an important classification of birds in 1827–28 which agreed remarkably—and quite independently—with the equally important classification of Nitzsch (1829).

Taxonomy. Classification, based upon the study of anatomy, and increasingly also upon geographical distribution and paleontology (and in the present century behaviour and parasites too), became an advancing science very rapidly after the developments of the early nineteenth century. The family tree of birds is not yet quite stable, but seldom now do new researchers saw off large branches and transpose them to another part of the tree, as happened quite often in the nineteenth century. Rather strikingly the publication of Darwin's *Origin of Species* (1859) made no great difference to ideas of the affinities of birds, which proceeded to develop until, under the care of the great scholar Alfred Newton [1829–1907], they reached a rational, if not final, arrangement about fifty years ago. Among those who served the cause of order well were C. W. L. Gloger (1834) who took much note of the works of Nitzsch in his arrangement of birds, realized the fundamental difference between swallows and swifts, and used the study of vocal muscles to distinguish between the singing and non-singing passerines. The great German anatomist J. P. Müller [1801–58] developed this further in 1845–47.

By halfway through the century the fundamental division

of birds into primitive flightless birds and the rest was established (Merrem). The cranes had been separated from the herons and allied with the rails, the gulls allied with the waders and separated from the petrels (l'Herminier). The vultures of the Old and New Worlds had been separated from each other and from the hawks, and the cuckoos, woodpeckers, parrots, nightjars, todies, hornbills, hoopoes and kingfishers grouped together (Nitzsch and others).

In 1859 C. E. Blanchard [1819–1900] separated the owls from the other birds of prey, which no previous taxonomist had unequivocally done; but his contempt for those who used external characters and *habits* to help in taxonomy looks silly nowadays, when behaviour-traits are commonly used to determine relationships[1].

In 1860 W. K. Parker [1823–90] published the first of a long series of papers on avian anatomy; it was on the 'difficult' bird *Balaeniceps rex*, the shoebill or whale-headed stork. This was typical of Parker's devotion to throwing light on the true relationships of hitherto unplaceable birds. By the time his life's work was done he had made great contributions to our knowledge of the relationship between the game-birds and the tinamous, the affinities of the kagu and the screamer, and on many passerines.

In 1863 Sir Richard Owen [1804–92] described what still remains the oldest known bird, *Archaeopteryx*, from the cretaceous slate of Solenhofen in Bavaria. Coming so soon after the publication of the *Origin of Species*, this evidence of a reptilian ancestor of birds was most exciting, and made a profound impression on biological thought. By 1867 Thomas Henry Huxley [1825–95] was able to work out a most complete and convincing classification of the entire class of birds. I quote from the account of Alfred Newton (1884), whose *Encyclopaedia Britannica* article on "Ornithology", a work of profound scholarship, will never be surpassed as a critical exposition of scientific history.

[1]A good example is the investigation of the geese and ducks by Delacour and Mayr (1945), who made some really important decisions about relationships, using habits as primary evidence. But ninety years ago such methods were scarcely foreseen, except by a few biologists of exceptional merit who were also field-naturalists, such as A. R. Wallace (1864).

> "In the spring of the year 1867 Professor Huxley, to the delight of an appreciative audience, delivered at the Royal College of Surgeons of England a course of lectures on birds. . . . Never assuredly had the subject been attacked with greater skill and power, or, since the days of Buffon, had Ornithology been set forth with greater eloquence. . . . Starting from the basis (which, undeniably true as it is, not a little shocked many of his ornithological hearers) 'that the phrase "Birds are greatly modified Reptiles" would hardly be an exaggerated expression of the closeness' of the resemblance between the two Classes . . . he drew in bold outline both their likeness and their differences, and then proceeded to inquire how the *Aves* could be most appropriately subdivided into Orders, Sub-orders, and Families."

Huxley's work has been described as 'the great palatal subdivision', for as his main evidence of relationships he chose to investigate the bones of the palate. One of the reasons for this, no doubt, was the fact that far more skulls were preserved at the Royal College of Surgeons than any other bird bones. This palatal study of his led him to make some very important groupings which have never since been overthrown—notably the grouping of pigeons, gulls and waders, their connexion with cranes, rails and cariamas and (in a more distant sense) the distinction between the game-birds and the tinamous and the relationship of the latter with the primitive ratites; the relationship between geese, ducks, screamers and flamingoes; and their not so distant relationship with storks, herons and ibises on the one hand and cormorants, frigate-birds, gannets and pelicans on the other. Perhaps Huxley's most important mistake was his allying of the tubenosed birds and divers with the gulls and auks. But this is quite an exception to the rule of his contribution, which threw light on almost every branch of the tree of birds.

This great classification of Huxley's was made almost purely by comparative anatomy; indeed, very largely by the comparative anatomy of the palate. In the world of ornithology it was the greatest triumph of a subject that, having perhaps now completed the chief part of its historical task, has become neglected if not actually unpopular.

Somewhat in contrast to the work of Huxley was that of C. J. Sundevall [1801–75], who published his *Tentamen* in 1872–74. Sundevall's work was based more on external than on internal characteristics, and Newton's criticism of it is a model of lucidity and restraint. Yet Sundevall managed to improve classification in some minor branches of the tree, particularly the passerines.

In 1874 there began to be published by the British Museum (Natural History) a famous document modestly called *Catalogue of the Birds in the British Museum.* The editorship and a great deal of the writing of fourteen of its twenty-seven volumes was entrusted to Richard Bowdler Sharpe [1847–1909]. To us, now, at the distance of half a century from its last volume, the great Museum Catalogue is still, in the words of Casey A. Wood (1931)

> "the most useful, as it is the most important, of all the works in any language on systematic ornithology. It furnishes a complete list of every bird, not only in the immense collections of the British Museum but of numerous public and private museums elsewhere. Distribution, synonomy, essential illustrations, both coloured prints and cuts in the text, are given, all combining to make this magnificent enterprise of the greatest value to every class of ornithological students".

Such a book will never be written again; for fifty years have added so much to knowledge of the world's birds that such a catalogue could not be written in 200 volumes. In fact, the modern catalogue has to *be* a catalogue—not (as the Museum Catalogue) a treatise. The catalogue of Peters (1931 onwards) which attempts to be no more, has been twenty years compiling without getting half way through the list of birds. More will be said about the British Museum Catalogue when later I touch on systematics. Meanwhile, it is enough to note that this great work provided a classification based quite radically on the work of Huxley and later investigators, and that while some of the groupings used might slightly surprise a modern ornithologist, he would encounter few such surprises and no outrages.

Further improvement in the classification of birds resulted

from the work of P. L. Sclater [1829–1913] who published a general arrangement of birds in 1880, largely based on Huxley, Parker and Garrod. A. H. Garrod [1846–79], whose life was cut short by tuberculosis, was a pupil of Newton's who, with another short-lived naturalist, W. A. Forbes [1855–83] made vast improvements in our knowledge of the classification of the passerines, particularly the anomalous forms such as the tyrants, pittas, ant-birds, broadbills and the lyre-bird. Sclater's work became for many years the standard classification for European ornithologists, though it was somewhat modified by Newton in 1884 and 1896.

In North America L. H. Stejneger [1851–1943] published an admirable and useful classification in 1884–85, which was followed in that continent and elsewhere also for many years. But perhaps the most important single contribution to improved taxonomy in the late nineteenth century was the great work of M. Fürbringer [1846–1920] which was published in Jena and Amsterdam in 1888. It is still the best original work on the minute anatomy of birds. Fürbringer proved the ratites to be unquestionably derived from fully-winged ancestors, a point of view which various workers have tried to upset subsequently but which stands firm. Fürbringer also found close relationship between the humming-birds and the swifts, and demonstrated the close alliance between the tubenoses and the penguins.

He worked in very close collaboration with Hans Gadow [1855–1927], who was also a close friend and collaborator of Newton. Gadow's admirable classification of 1892–93 is summarized in Newton's famous *Dictionary of Birds* of 1896, and it must be fairly said that any subsequent improvements to our knowledge of the avian family tree have been as the result of newly discovered fossils (and in regard to the position of those fossils); there has not been much consequent rearrangement of the branches. In some orders of birds there have been slight rearrangements at family and sub-family level. Thus since Gadow (1896) showed that the South American fossil bird *Phororhacos* was related to the cariamas and cranes, these groups have become progressively more close. The grebes and divers, which used to be thought to be closely related, are now (Stolpe, 1935) considered to be not so, yet it seems clear that both may be separately descended from tubenoses, which preserves them in the same branch of the family tree.

Lanham (1947) has shown that the tropic- and frigate-birds, which are primitive members of the Steganopodes (cormorants, gannets and pelicans) share a number of apparently primitive characters with the tubenoses, which may put their orders on the same rather than (as hitherto) separate but close branches of the avian tree.

Somewhat radical suggestions were made in 1953 by Jollie and Beecher. The former, on anatomical grounds so far unpublished, would identify four or five groups in the Falcones which, in his opinion, have reached their present taxonomic position entirely by convergence. He suggests that the Cathartidae or New World Vultures are close to the pelicans and tubenoses; that the apparent resemblance between the secretary-bird *Sagittarius* and the cariamas (in the Grues) is a real one; that the Accipitridae are an ancient (Eocene or earlier) group of unknown relationship; that the osprey ***Pandion*** may have diverged early from the accipitrid line or have reached its form by convergence; and that the Falconidae are quite distinct from, and probably later in origin than the Accipitridae.

Beecher's remarkable paper, based primarily on the muscle-anatomy of the head-region, concerns the Oscines or song-birds, the huge suborder of the Passeres that, with nearly four thousand known species, contains nearly half the world's birds: and is a radical and daring attempt, most lucidly argued, to construct both a family tree and classification based on true relationship. Some idea of his radical suggestions can be given by mentioning that he traces descent of the linnet-like finches from the tanagers; the buntings and New World 'sparrows' from the New World 'warblers', or wood-warblers; and the weavers and house-sparrows from the Old World fan-tailed warblers. Many other examples of convergence in form (e.g. nectar-probing) by entirely unrelated birds are given in this interesting analysis.

Apart from these modifications, the twentieth century has been primarily concerned with rearrangements of the class of birds which, so far as the tree of evolution has been concerned, has largely consisted of viewing it from different sides. Its real nature has remained on the whole unchanged. Rather clearly there develop two schools of arrangement. The *Zoological Society's Vertebrate List, Birds* (G. C. Low, 1929) and the *Handbook of British Birds* (Witherby and others 1938–41) are arranged on the old European system. This starts with the highest, believed to be the corvid passerines, and ends with the most primitive, the ostrich. The American system (Wetmore, 1934, 1951 and Peters, 1931 onwards) reverses the order and starts more logically, though perhaps not more biologically, with the ostrich. The order of its orders is not quite the mirror image of the European order, but in so far as it departs from it, it is more logical, and it is gratifying to be able to record that by an arrangement entered into at the International Ornithological Congress of 1950, European workers have agreed to go over to the 'Wetmore system'. This is an appropriate and entirely merited compliment to Alexander Wetmore, who was president of that Congress and for years was Secretary of the Smithsonian Institution, the world's greatest expert on fossil birds and avian evolution, the Grey Eminence of American ornithology.

I cannot close this chapter without some reference to the useful *History of Birds* of W. P. Pycraft (1910) and of E. Stresemann's contributions to Kükenthal's *Handbuch der Zoologie* (1927–34). Stresemann's contributions to classification are many, and in particular he throws much light on the families of the particularly difficult order Passeres. Mayr and Amadon's *Classification of Recent Birds* (1951) contains many useful commentaries on points of relationship that are still difficult.

Recently, new evidence on some of these knotty problems has come from the study of parasites. Thus, Hopkins (1942) shows that of the four known genera of Mallophaga or feather-lice, from flamingoes, three are strictly confined otherwise to Anseres, and the fourth also shows anserine affinities. Not one of the parasites suggests any close relationship between flamingoes and storks or herons. Hopkins also shows that the curious anomalous 'hammer-headed stork' *Scopus umbretta*, which is usually allied with the storks, has parasites belonging to the gull-wader order! He thinks, too, from the study of parasites, that the bustards have no close connexion, as has been thought, with the cranes, but may be closer to the game-birds. Rothschild and Clay (1952) show that the ostrich of Africa and the rheas of South America, which have been placed in separate orders, possess closely related species of a genus of Mallophaga (*Struthiolipeurus*) which is found on no other birds. But perhaps parasites point more clearly to the absence, rather than the presence of relationship. From their mallophagan parasites Clay (1951) would find the Falcones one true order; a view which, as we have seen, is disputed by Jollie (1953) on anatomical grounds. Jollie says that "the conclusion that Mallophaga are no more reliable in taxonomic matters than other single criteria is inescapable. It is possible that the nature of the food of these [four or five, he thinks] species groups has produced enough convergence in their physiological chemistry for mallophagans to be exchanged over shared carcases."

CHAPTER IV

SYSTEMATICS

ACCORDING to Mayr (1942) the terms taxonomy and systematics can be considered to mean approximately the same thing. "In America", he writes, "the term taxonomy seems to be preferred, in the rest of the world the term systematics seems to be more widely used". Mayr perhaps overlooks the fact that in his time, very much owing to the work of 'new' systematists like himself, the two words have come to mean the same thing as a result of the improvement of our understanding of evolution. When systematics 'began', which can be said to be with the publication of the tenth edition of the *Systema Naturae* of Karl von Linné (Carolus Linnaeus) in 1758, it was a very different thing from taxonomy. For taxonomy means the study of the true relations of the family tree of past and present living things, whereas systematics is the study of the best practical way to arrange them. Ideally, of course, taxonomy and systematics should thus be identical, and they would be so if it were possible for us to know the place and time of every branch in the tree of evolution and to create a great model of it in three dimensions upon which we could grow, as branches and leaves, all past and present species. But the systematist, however clever he is, however subtle a student of evolution, however well-trained a biologist, has to confine the results of his work in museum cabinets and on pages of paper.

It is obviously impossible to produce a truly consecutive list of orders, families, genera and species of birds, which, while it is consecutive, at the same time depicts the true biological relationship of its members. The 'new' systematics (an expression which has been in use for nearly twenty years) has been concerned with ensuring that the practical everyday arrangement of species reflects their true biological relationships as much as possible. As Mayr points out, the old system was characterized by the central position of the species, which was almost always defined on purely morphological grounds. These species were described from one or a few specimens. The describer was usually obsessed with technical questions of nomenclature, and had invented a private language in which to talk about them to his fellow systematists. "The major problems are those of a cataloguer or bibliographer, rather than those of a biologist."

Mayr then goes on to characterize the new systematics as one in which the importance of the species is reduced, and where, rather than an individual specimen, a whole series or even 'population' of specimens has become the basic systematic unit. The actual definition of species changes to become a biological one, still highly complex and difficult, largely because it has to take so many different factors—ecological, geographical, genetic and so on—into consideration.

The convergence of taxonomy and systematics to the situation in which they can be thought to mean the same thing is a process which has taken 200 years, though the terms have not, of course, been in use for that whole time.

The father of modern systematics set out consciously and deliberately to invent a system of nature. The first edition of his *Systema Naturae* was published in 1735. Linnaeus was born in Sweden in 1707 and went as a student first of all to the University of Lund and then in 1728 to that of Uppsala. He studied under the famous botanist Olov Rudbeck the Younger. Although Rudbeck was primarily a botanist, he gave several lectures on birds, which were attended by Linnaeus. For most of his life Linnaeus was associated with the University of Uppsala, first as pupil, then as teacher, finally as professor. He made that northern city into the hub of the world's natural history for nearly half a century. His passion was botany, but he wrote almost as much on zoology, his obsession being with description, in which he showed an economy of style and crispness of definition which were unprecedented, and with the naming and ordering of living things in a systematic way. His great invention was the binomial system of nomenclature, which is now adopted throughout the world, in which each plant and each animal bears a scientific name, usually, though not necessarily always, of classical derivation or in classical style. The name of the species is of two words which make it only when together (the specific name is *not* the second of the two names, as is generally thought; this is simply an adjectival description of the first). The first word is the genus, and there may be several species belonging to one genus, distinguished by the use of different adjectives (second words) describing it.

Upon the foundation of the Linnaean system of nomenclature the whole of modern systematics is built, and no post-Linnaean innovation is more than a refinement. In most branches of the animal kingdom and a few of plants a third name has now been introduced to be descriptive of geographical races of the same species. These are known as subspecies. Various rules have been introduced and agreed upon, such as the citation of the authority who gave the name. This citation must always be used when the name is formally set out. There is an International Commission on Zoological Nomenclature, which is steadily improving the international rules of nomenclature, all of which are based on the Linnaean system.

Linnaeus's tenth edition of the *System of Nature*, of 1758, compiled when he was at the height of his powers, is the first book about living things in which the concept of species is adopted and unequivocally stated.

To us, who have lived all our lives with the species concept, the reality conveyed by the word species is a definite one; we understand in a practical way what a species is. For instance, among the 426 species of the list of British birds published by the British Ornithologists' Union in January 1953 not one's status as a species is questioned by any ornithologist, as far as I can tell. The most controversial case is whether the bean- and pink-footed geese are well-marked geographical races of the same species or whether they should be regarded as separate species whose breeding range nowhere overlaps.

We have already discussed the rise of the list of British birds in a previous chapter, in order to demonstrate the history of the development of an interest in bird-watching and ornithology. An equally interesting idea of the rise of ornithology can be gained from a study of new species since the tenth edition of the *System of Nature*. In that book Linnaeus described 564 species of birds from all parts of the world. Well over 500 of these species still pass muster on Linnaeus's original description, there being no possibility of their being confused with any other. For the rest, it is not possible now to be sure just what species Linnaeus was describing; in some cases he embraced in one description more than one species; in others he gave separate descriptions as separate species of specimens of what is now known to be one. The small number of such differences of judgement between Linnaeus and the systematists of today is striking. It means that if he had lived today he would probably have been, still, the best systematist alive.

In an interesting article Mayr (1946) has discussed the number of species of birds in the world. He shows that, a century and a half after Linnaeus, no less than 2,810 genera of 18,937 species were accepted by R. Bowdler Sharpe (1899–1909).

In the days of Sharpe the concept of subspecies was in its infancy, and there resulted a quite understandable confusion of races and true species. Sharpe, careful and excellent ornithologist as he was, actually listed more than twice as many species as he would have done if he had lived eighty years later and been concerned with biological concepts of species. Since 1909 about 400 new true species have been described, acceptable by the most rigorous standards. But the fact remains that Sharpe's great list embraced only about 8,200, not nearly 19,000 species, as he thought. By the end of 1945 Mayr estimated that the number of valid subspecies described amounted to about 28,500, half as many again as the 'forms' of Sharpe. The number of true species accepted was 8,616. Mayr pointed out that the number of valid subspecies described was then increasing annually by about 200, but he doubted whether in the entire world even as many as 100 new species remained to be discovered. In judging this he did not, I believe, take account of the inevitability of new species presently appearing in the course of evolution. The change of geographical subspecies into full species is achieved by the crossing of a threshold which is still rather undefinable, and we still know little about the *rate* of speciation, except that it probably differs in different groups.

How did the world progress from the 500 bird species of Linnaeus to the 8,600 of Mayr in a world almost exhausted of ornithological novelty? The answer is in what remained of the eighteenth century, by a series of leaps, each coinciding with the publication of an important catalogue of descriptions; and in the nineteenth century, with the rise of scientific journals and scientific accounts of voyages, much more evenly and smoothly (though not uniformly); and in the twentieth century with returns which, at least as regards full species, were obviously diminishing.

The association, in all the textbooks, of the names of the describers and the dates of publication of their descriptions (with the names they gave to the birds) give us all an ever-present reminder, as we peruse the books, of the powerful systematists of the past. Though Linnaeus described species from all quarters of the world, his attention was naturally first of all for the birds of Sweden (of which most are also found in Britain); so that in any list of British birds we are continually reminded of Linnaeus or Linné or Linn, or L., usually with the date 1758, sometimes with the date 1766, that of the twelfth edition of his *System of Nature*, occasionally with other dates such as 1761 (the second edition of his *Fauna of Sweden*).

Under the rules of nomenclature each living thing bears the first name which was given to it under the binomial system with a reasonable and sufficient reason in a work actually published (printed is not enough) since 1758. This has stimulated many students and scholars to delve frantically into the past; and has resulted in the unearthing of works of the utmost rarity and obscurity. While this has taken many scholars a great deal of time, it has resulted at last in a measure of stability, though many people now think that some names which are still controversial should be 'frozen' more often by decisions of the International Commission on Zoological Nomenclature, which has the power to make them. Certainly a lot has been discovered to please, interest and amuse the bibliophile. One of the interesting results of the system of priority is that certain naturalists have come into a prominence perhaps greater than that which they really deserve, simply because they took species discovered by others and bestowed on them the first binomial names. In cases of this kind it is the man who gives the name and not the discoverer of the species who is quoted as the author of the name. To give an example, the collector Giovanni Antonio Scopoli describes in his *Deliciae Florae et Faunae* (1786–88) quite a number of species discovered by the much more important though somewhat inaccurate naturalist Pierre Sonnerat, on his voyages to New Guinea, the Philippines, China and the East Indies. The original *Table des Planches enluminées* of Pieter Boddaert was published in Utrecht in 1783 and is one of the rarest early descriptive books on the Linnaean system. Boddaert only printed fifty copies of this book, of which two are in Britain. The work is a list of the famous *Planches enluminées* of E. L. Daubenton, who, instigated by the great naturalist Buffon, began in 1765 the publication of a series of 1,008 hand-coloured engravings drawn by Martinet, of which 973 are of birds. These appeared gradually without any text other than their vernacular names, as an appendix to the Comte de Buffon's *Histoire naturelle générale* (Daubenton, 1765–81).

Boddaert's table to the *Planches enluminées* is full of typographical errors and exhibits no originality except for a considerable number of new binomial names. The book is certainly one of the 'headaches' of the priority system, and its errors, in the words of J. T. Zimmer (1926), a notable modern systematist, "present numerous problems which have yet to be solved". So Boddaert is immortalized for other people's work. Johann Friedrich Gmelin, editor of the important thirteenth edition of Linnaeus's *Systema Naturae*, published in 1788–93, has been responsible for the introduction of more binomial names of birds than any other worker since Linnaeus. Alfred Newton (1884) is worth quoting on the subject of Gmelin.

> "Gmelin", Newton writes, "commenced what he called the Thirteenth Edition of the celebrated *Systema Naturae*, which obtained so wide a circulation that, in the comparative rarity of the original, the additions of this editor have been very frequently quoted, even by expert naturalists, as though they were the work of the author himself. Gmelin availed himself of every publication he could, but he perhaps found

> his richest booty in the labours of Latham, neatly condensing his English descriptions into Latin diagnoses, and bestowing on them binomial names. Hence it is that Gmelin appears as the authority for so much of the nomenclature now in use. He took many liberties with the details of Linnaeus's work."

It is easy to understand Newton's solicitude for that very much more important naturalist, John Latham, who began his *General Synopsis of Birds* in 1781. Latham's many years of work were inspired and fed by the opening up of new trade routes, the exploration of the farthest parts of the world and the consequent new mass of ornithological material, especially from Australia and the Pacific. He found abundant material for his researches in the recently established British Museum and in another museum belonging to Sir Ashton Lever. His work was not always accurate and, much more often than Linnaeus, did he make the mistake of describing the same species twice, in his vast *General Synopsis of Birds*, with coloured plates. But not until 1790, with his *Index Ornithologicus* did Latham publish his own systematic catalogue of the birds in the world with Linnaeus's binomial system. With this Latham's contributions to systematics, excellent though they were, may be said to have ended, for although he lived till 1837 and the age of ninety-seven, and published a general history of birds in eleven volumes in Winchester in 1821–28, "his defects as a compiler, which had been manifest before, rather increased with age and the consequences were not happy" (Newton, 1884).

To show the rise of our systematic knowledge of the world's birds I have taken the first volume of the *Check-list of Birds of the World* by the late J. L. Peters (1931), which contains a list of all the acceptable species whose names were published up to 31st December, 1930, of living or recently extinct ratites, tinamous, penguins, divers, grebes, tubenoses, steganopodes, herons, anseres and diurnal birds of prey. By rejecting a few now known to be synonyms, I have brought the list to 847 species, though it should perhaps be a little less, since various other species considered separate by Peters are not so regarded by most recent authorities. But in comparison with the total, these can make little difference. Of the 847 described by 1930, three-quarters (636) had been described by 1851, half (430) by 1824, a quarter (214) by 1788 and ninety-five, or rather under one-eighth, by Linnaeus. Actually, of the birds of the world, Linnaeus only described one-sixteenth, but the orders in this first volume are of rather more familiar and well-known kinds, which thus tended to be described earlier than (for instance) the small passerines of the tropical forests. Unfortunately there is as yet no list of the world's passerines. Such a thing only began with the latest volume (the seventh) of Peters' list (1951).

In this reckoning, of course, I have considered only the first descriptions of the species at present recognized, not the number of species recognized at the time the descriptions were published.

It can be seen that three-quarters of these rather familiar orders of birds were 'known' by the middle of the nineteenth century. Some members of the more elusive families were not recognized for much longer. Even the largest and most spectacular of the passerines of the tropical forests—the birds of paradise—were not discovered for a very long time, and only about half of the present known species (forty-three) of birds of paradise had been described by 1860. The latest to be discovered was *Astrapia mayeri*, the ribbon-tailed bird of paradise, described by C. R. Stonor (1939) from two central tail-feathers.

An analysis of the describers of the 426 species of the British List of 1952 (to 1950) is illuminative of the power of Linnaeus in the rise of systematics. Linnaeus was the first describer of 241 of the British species—216 (or over half) of them in his *Systema Naturae* of 1758, in which somewhat over 500 accepted species were first described. Of the rest, thirty-four were first described by P. S. Pallas between 1764 and 1811, fifteen by J. F. Gmelin between 1771 and 1789, fifteen by C. J. Temminck between 1815 and 1840, fourteen by L. J. P. Vieillot between 1816 and 1819, ten by G. A. Scopoli between 1769 and 1786, nine by J. Latham between 1771 and 1801, seven by J. M. Bechstein between 1793 and 1812, and six by Bishop E. L. Pontoppidan in 1763. Others who first described more than two British species are Brünnich and Boddaert (five each), and Forster, Leisler, Lichtenstein and

Blyth (three each). There are eight authorities who have described two each, and thirty-seven who have described one. The last year in which any species on the present British List was first described was 1863, when Schlegel described *Procellaria* (now *Bulweria*) *neglecta*, the Kermadec petrel, Radde *Sylvia* (now *Phylloscopus*) *schwarzi*, the bush-warbler that now bears Radde's name, and Swinhoe *Anthus gustavi*, the Petchora pipit. All these are of course stragglers to Britain; the first two have only been recorded once: the regular-breeding or common visiting British species were all first described in the eighteenth century, except for sixteen[1].

The mainstay of the ornithology and systematics of today is so obviously the scientific bird journal (with its supporting institute or society) that it is difficult to imagine that about ninety per cent of the species of birds were first described in books, and that with a few minor exceptions there were no journals at all a hundred years ago.

Actually the world's first bird journal was published in 1824. C. L. Brehm's *Ornis* only lasted till 1827; it was published in Jena, and mostly written by its editor, with many useful reviews of contemporary books. For nearly twenty years after the premature and short flourishing of *Ornis*, there was nothing; then from 1846 to 1849 another short-lived German periodical, *Rhea*, lived for two volumes under the editorship of F. A. L. Thienemann. In 1848 came Sir William Jardine's *Contributions to Ornithology*; the last of its five volumes was published in 1853. *Naumannia*, the organ of the newly formed *Deutsche*

[1]Leach's petrel, *Oceanodroma leucorhoa* (Vieillot, 1817), the great shearwater, *Procellaria gravis* O'Reilly, 1818, the bean- and pink-footed geese, *Anser arvensis* Brehm, 1831, the barnacle-goose, *Branta leucopsis* (Bechstein, 1803), Bewick's swan, *Cygnus bewickii* Yarrell, 1830, the little stint, *Calidris minuta* (Leisler, 1812), Temminck's stint, *C. temminckii* (Leisler, 1812), the pomarine skua, *Stercorarius pomarinus* (Temminck, 1815), the long-tailed skua, *S. longicaudus* Vieillot, 1819, the Iceland gull, *Larus glaucoides* Meyer, 1822, the arctic tern, *Sterna macrura* Naumann, 1819, the roseate tern, *S. dougallii* Montagu, 1813, the nightingale, *Luscinia megarhynchos* Brehm, 1831, the reed-warbler, *Acrocephalus scirpaceus* (Hermann, 1804), the chiffchaff, *Phylloscopus collybita* (Vieillot, 1817), and the firecrest, *Regulus ignicapillus* (Temminck, 1820). The last to be described, the nightingale, had previously been confused with the thrush-nightingale, *Luscinia luscinia* (Linnaeus, 1758); and a glance at this list shows that the rest are also species which are all liable to be confused with others, and were by the systematists of old.

ornithologische Gesellschaft, was first published in 1851 under the editorship of Edward Baldamus and in 1860 was incorporated in the *Journal für Ornithologie*. This started in 1853 and with the exception of a short intermission in and after the Second World War, has been regularly published ever since; it is the oldest current bird journal. The *Ibis*, the journal of the British Ornithologists' Union, has never missed a number since it was first published in 1859; it is the second oldest current journal. The third oldest is the *Auk*, the periodical of the American Ornithologists' Union, which began in 1876 as the *Bulletin of the Nuttall Ornithological Club*, and assumed its present name in 1884.

The influence of the *Journal of Ornithology*, *Ibis* and *Auk* on ornithology has been prodigious. Wood (1931, p. 413) suggests that they comprise "three-fourths of the original and essential literature on the subject of birds in their various relations. From these three sources have been drawn the great majority of references to and illustrations of new discoveries in avian life". Certainly throughout the career of these three great journals they have contained the best original papers of their day, and have continually, in reviews and abstracts, kept their readers abreast of all contemporary publications and researches. They still dominate ornithology, and there seems no reason why they should not continue to do so; even though there are now many rivals and competitors. It seems likely, from a survey of the titles of journals listed in the *World List of Scientific Periodicals*, by various library catalogues, and particularly by Strong (1939), that, excluding journals which are mimeographed, purely annual reports, or purely devoted to poultry or cage-birds, there have been at one time or another about 250 different ornithological journals. Some important ones, mostly still current, with the dates of their foundation (occasionally under different names) are:—

1876: *Ornithologische Monatsschrift* (Germany).
1885: Proceedings of International Ornithological Congresses (*Ornis*).
1889: *Wilson Bulletin* (U.S.A.).
1893: *Bulletin of the British Ornithologists' Club.*

1893: *Ornithologische Monatsberichte* (Germany).
1894: *Aquila* (Hungary), *Avicultural Magazine* (Great Britain).
1897: *Verhandlungen der ornithologischen Gesellschaft in Bayern* (Germany).
1899: *Condor* (U.S.A.).
1900: *Pacific Coast Avifauna* (U.S.A.).
1902: *Bird Notes* (G.B.), *Emu* (Australia), *Ornithologische Beobachter* (Switzerland).
1906: *Dansk ornithologisk Forenings Tidsskrift* (Denmark).
1907: *British Birds.*
1909: *L'Oiseau* (France).
1911: *Rivista Italiana di Ornitologia.*
1912: *Ardea* (Holland).
1913: *Nos Oiseaux* (Switzerland).
1914: *South Australian Ornithologist.*
1915: *Tori* (Japan).
1920: *Murrelet* (U.S.A.), *Norsk ornithologisk Tidsskrift* (Norway).
1921: *Oologists' Record* (G.B.).
1925: *Beiträge zur Fortpflanzungsbiologie der Vögel* (Germany), *Bird-Banding* (U.S.A.).
1926: *Bulletin of the British Oologists' Association* (G.B.).
1928: *Limosa* (Holland).
1929: *Alauda* (France), *Kócsag* (Hungary).
1930: *Ostrich* (S. Africa), *Vogelzug* (Germany).
1935: *Bulletin of the British Trust for Ornithology* (now *Bird Study*).
1942: *Vår Fågelvärld* (Sweden).
1943: *Notornis* (New Zealand).

From this list the reader will guess that practically only professional ornithologists (and not all of them) can keep up with modern ornithological literature. An enormous amount is published in museum, ecological and zoological journals all over the world, quite apart from the purely ornithological journals. It is probable that the number of words on birds annually published exceeds 10 million. Nobody could possibly read all.

A good way of understanding how thorough is present

knowledge of avian systematics is to study the descriptions of valid new species during the last thirty years. Since 1920 practically all descriptions of new species of birds in the world literature have been most carefully criticized and assessed by Meise (1938, 1938b), Zimmer and Mayr (1943) and Mayr (1946). Since Mayr's summary of the number of species of birds on 30th December, 1945, 8,616 species, I have noted further new species by exploring the *Zoological Record*, though I have possibly not been able thus to find them all, far less examine them with the critical accuracy of Meise or Mayr.

In 1946–1950 inclusive, at least nineteen new species were described, though one of these is certainly inadmissible and two highly doubtful. Moreover the new count of Mayr and Amadon (1951), probably in December 1950, gives only 8,590 world species, in spite of these additions; showing that, in a new appreciation of the status of all world species, the effect of new additions can be quite swamped by the effects of even slight changes in judgement. We can expect the list of world bird species to continue to creep up at the rate of two or three a year for quite a time. As we have mentioned, in 1946 Mayr doubted whether as many as 100 new species remained to be discovered in the entire world. The world total of bird species is thus liable to be altered far more radically by changes in systematists' opinion of the nature of species, than by the addition of any new ones.

During the period 1920–1934 about 170 good species according to Meise (1938) and about 150 according to Zimmer and Mayr (1943) were described (about ten or a dozen a year). In 1935 twelve new species were described, in 1936 five, of which at least one is invalid, in 1937 seventeen, of which at least two are invalid, two doubtful, and one described on a single specimen. Thenceforward the figures run:—1938, fifteen, of which nine are probably valid; 1939, nineteen, of which six are probably valid; 1941, nine, of which two are probably valid.[1]

[1]In summarizing the 1938–41 situation, Zimmer and Mayr (1943) state that twenty-nine Old World and twenty-three (they record twenty-four by a mistake) New World species were described. Actually the Old World figure was thirty-one, as they had not then seen two valid new species described by Hachisuka. Apart from these two, they assessed thirteen Old

Since 1942 the number of new species described run:—1942, five; 1943, eight; 1944, three; 1945, three; 1946, seven; 1947, three; 1948, two; 1949, four; 1950, five; 1951, five; 1952, three. Some of these are certainly invalid, but the proportion of invalid species has been lower in this last decade, owing to the improvement of critical standards.

As might be expected, most of the new species since 1920 have been described from the tropical passerine fauna. But there are still some remarkable 'finds' from time to time. Thus, as late as 1949, R. C. Murphy described a new species of gadfly-petrel from the Pacific, *Pterodroma ultima*, which is not only large and striking but also abundant. In 1948 Zimmer described a new species of trogon from Colombia. In 1943 Stresemann described a new species of crow from Queensland, Australia, in 1942 Benson a new swallow from Abyssinia, in 1940 Rand a new bower-bird from Dutch New Guinea, which he placed in a new genus. In 1939, besides Stonor's discovery of the ribbon-tailed bird of paradise, Gyldenstolpe described a remarkable new shrike from Northern Annam, and Oscar Neumann a new species of owl from the Celebes.

I have so far singled out deliberately those new species which have been somewhat of a nine days' wonder. There has been nothing quite as spectacular as the ribbon-tailed bird of paradise since 1939; but in the early 'thirties there were a number of quite remarkable discoveries, including a new owl, a new pheasant, a new tinamou, a new drongo, a new pipit and a new starling. One of the most interesting events was the discovery of *Zavattariornis stresemanni*, a corvid or sturnid bird of an entirely new genus and species from southern Abyssinia, first described by Moltoni in 1938. Actually the affinities of this bird are rather uncertain, for its Mallophaga suggest a relation with the Sturnidae, as do some of its habits, while its anatomy, according to Lowe (1949) may be far enough from both Corvidae and Sturnidae to warrant a separate family.

World and eleven New World species as valid. Actually, the figure should probably have been fourteen and ten, as *Grallaria nattereri* was described by Pinto in 1937, and in any case was later showed by Peters (1951, p. 273) to be a synonym of *G. ochroleuca*, while *Astrapia mayeri* of Stonor (1939), rejected by Zimmer and Mayr, has proved, as we have seen, a valid, indeed remarkable, species.

When Moltoni first described the Abyssinian pie it was only known from two specimens, but Benson (1942, 1946) visited the type locality, Yavello, in 1941 and 1942, where he found it numerous, though restricted to an extraordinarily small area. He comments that if it had not been for the restriction of its distribution, so conspicuous a bird would have been discovered long before.

Another interesting discovery was the green-faced parrot-finch of Hachisuka and Delacour (1937) from Manila in the Philippines. Actually the first records of the bird date back to 1920, when W. Richmond suspected they might belong to a new form. Early in 1935 Hachisuka (1937) found in the establishment of a bird-dealer in Los Angeles a number of these finches. Indeed, they even had a trade name, Luzon finch, and Hachisuka found they had been exported from Manila to San Francisco in hundreds and distributed in many aviaries and even zoos, which had exhibited them for some time without knowing their identity. It is interesting that a fairly common cage bird should be innominate as late as 1937.

But perhaps the most remarkable species discovery of the twentieth century was the recognition of the Congo peacock, *Afropavo congensis* Chapin, 1936. In 1913 J. B. Chapin's attention was drawn to a peculiar feather in the hat of a native of the Ituri forest of the Congo. It had apparently come from some gallinaceous bird larger than a guinea-fowl. He brought this secondary quill-feather home, showed it to many other ornithologists, but could find no bird from which it might have come. In the following year the museum of the Belgian Congo acquired from another Belgian museum some mounted birds which had mostly been collected before 1900. When these skins reached the museum (at Tervueren, Belgium) two of them, which bore a slight resemblance to the peacock and which were already labelled *Pavo cristatus*, were put on top of a cabinet in a room not used by ornithologists. "There they stood in seclusion until early August 1936", writes Chapin. "At that time I happened into the room, noticed the birds, and stood dumbfounded. . . . Here at last, I realized, was the bird from which my feather had come." Later Chapin was told by a Belgian friend of a strange bird he had eaten in the Congo forest in 1930: this man drew a sketch of its head which showed that he meant the same bird. It had become clear that in the Congo forest lived a large member of the pheasant family resembling a peacock. Until then, no peacock-like members of the pheasant family were known in Africa.

The last new species to be described from anywhere in the Holarctic region is the putative *Dendroica potomac* Haller, 1940, a new wood-warbler from West Virginia, of all places. Quite rightly, Sutton's warbler, as it is now called, must be placed as a doubtful species of possible hybrid origin (Zimmer and Mayr, 1943, p. 254). The new bird was first seen by K. W. Haller and J. L. Poland on 30th May, 1939 in mixed oak-pine woodland in the eastern panhandle of West Virginia. A very good description of later observations of this interesting form is given by Peterson (1950, pp. 184–87). The fact that several reliable observers have seen living birds and that their characters appear to be constant is some evidence that the form may not, after all, be a hybrid between the yellow-throated and Parula warblers, *Dendroica dominica* and *Compsothlypis americana* (Sutton, 1942; Cockrum, 1952). If *Dendroica potomac* eventually proves to be an unacceptable species, then the last North American new species to be described is *Ammospiza mirabilis* (Howell, 1919), which is restricted to a few square miles of the coastal marshes near Cape Sable, Florida, the Cape Sable seaside sparrow. As Peterson notes, ornithologists had predicted that this would be the last new species to be found in the New World north of the Mexican border, and this may be true.

CHAPTER V

GEOGRAPHICAL DISTRIBUTION

WE have seen in previous chapters that about ninety per cent of the birds in the world had been discovered and described by 1850, a hundred years ago. This situation, of course, could not have been reached without the fine exploration of every quarter of the globe, every important tributary of the great tropical rivers, every remote mountain range or desert. The discovery of the last ten per cent of species, which has been almost completed in the last hundred years (it is very unlikely that there are a hundred left to discover) has not thrown any important light on relationships at the level of families and orders. Knowledge of the families of birds eighty to a hundred years ago was detailed enough for relatively advanced exercises in geographical correlation. It was in the *annus mirabilis* 1858, that on the first of July, in a quiet meeting in the rooms of the Linnean Society of London, Darwin and Wallace together first laid before the scientific world those views that were to shake it to the foundations. In the same year, in the *Zoological Journal* and *Proceedings* of the same Society, P. L. Sclater published a paper. 'On the general geographical distribution of the members of the class Aves'. This communication, overshadowed as it was by the mighty ideas of Darwin and Wallace, represented nevertheless a crystallization of thought, a new synthesis that possessed the same attributes of boldness and scholarship, a buttress for the evolution theory.

The first person who seems to have had a reasonable theory about the geographical distribution of animals was Buffon, though, as Newton (1884) points out, "this theory was scarcely touched in the ornithological portion of his work [1770–83], and has since proved to be not in accordance with facts". In another place (1896, p. 311), Newton writes of Sclater "though here and there in the writings of his predecessors truths are doubtless apparent, it is certain that no one had hitherto taken the question seriously in hand, and that such truths as had been reached were rather by favour of fortune than by application of knowledge".

It must be stressed that Sclater developed his ideas working with birds alone, with what would now be regarded as a crude knowledge of the true relationships of living families and orders of birds, and without the background of the theory of organic evolution (though he certainly provided background for this theory). It is a very striking thing that with only one important and only a few minor amendments, his suggested zoogeographical regions of the world still stand after the lapse of nearly a century.

Working simply on the families and orders of birds, and their occupation of the world, Sclater divided the land surface of the globe into six regions, in each of which he found an avifauna differing markedly from that of any other—though the differences were not always equally great. These regions were (a) the Palearctic region: Europe and Asia north of India and China, and Africa north of the Atlas mountains, (b) the Ethiopian region: the rest of Africa, Madagascar and southern Arabia, (c) the Oriental region: India and China, the East Indies to Borneo and Java, (d) the Australasian region: including Australia and New Zealand and their islands, the East Indies from Celebes and Lombok eastwards, and nearly all the Pacific islands, (e) the Nearctic region: North America, including Greenland and northern Mexico, (f) the Neotropical region: southern Mexico, Central and South America, and the Caribbean islands.

In the same year, 1858, Günther showed that the geographical distribution of reptiles agreed very closely with that of birds, though the relationships of snakes might carry the Oriental region as far as Japan. Ten years later, in 1868, Thomas Henry Huxley suggested an entirely different arrangement, based on the study of both birds and mammals, of the world's zoogeography, into a northern grouping which he called Arctogea, embracing all the world including Africa, except a southern grouping, Neogea, which included Australasia from Lombok and Celebes east through the Pacific to embrace the Neotropical region of Sclater. Huxley, incidentally, was the first to give the name 'Wallace's line' to the famous and very clear-cut division of life across the East Indies which, between the islands of Bali and Lombok, marks the most radical faunal change to be found over any twenty miles of the earth's surface.

It was his field study of this boundary between the Oriental and Australian regions that first led Alfred Russel Wallace (1859) to develop his notions of geographical distribution. There was always a long, though by no means ponderous, period of gestation and germination between the sowing of the seed of one of Wallace's ideas and the bearing of its fruit. If this had not been so he might perhaps have been given wider credit for his share of the theory of organic evolution, though it was no fault of Darwin's that his contribution tends to be overlooked. However, Wallace has never lacked credit for his views on geographical distribution, and indeed, his great two-volume work of 1876 is often recommended to students as the first full general promulgation of the ideas already most adequately presented by Sclater and Huxley. Wallace's *Geographical distribution of animals* is a most wonderful treatise, illuminated by multitudinous examples from the class of birds.

Wallace rejected the regions of Huxley in his book and returned with minor modifications to the six regions first proposed by Sclater; and indeed these remain the basis of our classification of the zoogeographical regions of the land today, with two important modifications—taken from Huxley. These are, his suggestions that the affinities between the Nearctic and Palearctic regions of Sclater are too great to permit them to be divided, as principal regions; but that New Zealand and its satellite islands are sufficiently distinct in their faunas from the rest of Australasia to be entitled to full regional status. The two northern regions are therefore united as the Holarctic region, and the Nearctic and Palearctic become sub-regions.

Most of the boundaries between the great regions are now very well understood. Between the sub-regions Nearctic and Palearctic they are not so well marked, which is in keeping with their sub-regional status, but it would appear that Alaska and East Greenland should be in the Palearctic region, judging by their bird fauna particularly, and Africa north of the Atlas remains in the Palearctic, as do the Canary Islands, Madeira and the Azores; and there are some reasons for placing even the Cape Verde Islands in the Palearctic. South-west Arabia, as Philby (1937) most recently points out, is certainly in the Ethiopian region, but the rest of Arabia and Persia are certainly in the Palearctic, the Oriental region not starting until Baluchistan, and being separated from the Palearctic by the Himalayas and the Tibetan plateau, the valley of the Hwang-Ho being Oriental but Korea and Japan Palearctic. Into the Indian Ocean the Oriental region extends as far as the Chagos Archipelago, but the Seychelles and Mascarene Islands are in the Ethiopian region. The line through the East Indies runs boldly between Bali and Lombok, but further north it is more equivocally drawn, there still being some doubt as to whether it lies east or west of Celebes, whose intermediate position has been thoroughly analysed by Stresemann (1939). The Philippines are certainly Oriental as are the Ryukyu Islands, south of Japan. In the Pacific, all islands except those fairly near the American coast are Australasian, including the Hawaiian islands, but the Galápagos islands belong to the Neotropical region, as do Mexico and Lower California. In the North Atlantic Bermuda has a fauna intermediate between Nearctic and Neotropical.

All orders of birds, except certain sea-birds like penguins and tubenoses, can be shown to 'belong' to one or more of these zoogeographical regions. Some, of course, are found in all of them, but there is no order which is found, as it were, in half one and in half another. And even some sea-bird orders or suborders are grouped according to some primary region. Thus, the Steganopodes or pelican-like birds probably originated in the Oriental region, and the gull-wader order in the Holarctic region.

Quite a number of the more primitive orders are entirely restricted to one principal region: the ostrich to the Ethiopian region, the kiwis to New Zealand, the cassowaries and emus to the Australasian region, the rheas and the tinamous to the

Neotropical, the loons to the Holarctic, the mouse-birds to Africa. No order (apart from sea-birds) is found only in two regions. The trogons are found only in three (Neotropical, Ethiopian and Oriental). Orders that are found only in four are flamingoes (not Australasian or New Zealand) and the Pici (woodpeckers) which only *just* reach the Australasian region and not New Zealand. Orders that are found in all but one region are the parrots (not Holarctic), and the Caprimulgi (goatsucker group) and Macrochires (swifts, etc.) which do not reach New Zealand. All the other orders—grebes, Steganopodes, falcon-eagles, heron-storks, ducks-and-geese, game-birds, cuckoos, crane-rails, gull-waders, pigeons, owls, rollers, etc. and passerines are cosmopolitan.

At family level the world distribution of birds is often of interest. Families restricted to one region are often primitive or specialized; examples from the Ethiopian region only are the secretary-bird (Sagittariidae in Falcones), hammer-head (Scopidae in Gressores), turacos (Musophagidae in Cuculi), roatelos (Madagascar only, Mesoenatidae in Grues), extinct dodos (Mascarene Islands only, Raphidae in Columbae), honey-guides (one Indian, Indicatoridae in Pici), and in the order Passeres the helmet-shrikes (one Indian, Prionopidae) philepittas (Madagascar only, Philepittidae) and tangas (Madagascar only, Tangidae).

Examples from the Neotropical region only are screamers (Anhimidae in Anseres), guans and the hoatzin (Cracidae and Opisthocomidae in Galli), cariamas, trumpeters and the sun-bittern (Cariamidae, Psophidae and Eurypygidae in Grues), seed-snipe (Thinocoridae in Laro-Limicolae), potoos and the oil-bird (Nyctibiidae and Steatornithidae in Caprimulgi), motmots and todies (Momotidae and Todidae in Coraciae), humming-birds (Trochilidae in Macrochires—a few in Nearctic), puffbirds, jacamars and toucans (Bucconidae, Galbulidae and Ramphastidae in Pici); and in the order Passeres the tapaculos (Rhinocryptidae), ant-pipits (Conopophagidae), ant-birds (Formicariidae), ovenbird-woodhewers (Furnariidae), tyrants (Tyrannidae), manakins (Pipridae), cotingas (Cotingidae), plant-cutters (Phytotomidae) and the swallow-tanager (Tersinidae). There are more specialized groups of birds in Central and South America than anywhere else in the world.

Examples from the Australasian region only are brush-turkeys (Megapodiidae in Galli), the kagu (Rhynochetidae in Grues), the owlet-frogmouths (Aegothelidae in Caprimulgi) and in the Passeres the lyre-birds (Menuridae), scrub-birds (Atrichornithidae), cuckoo-shrikes (Campephagidae), flower-peckers (Dicaeidae), honey-eaters (Meliphagidae: one in Africa), Hawaiian honey-creepers (Drepaniidae), bell-magpies (Cracticidae), magpie-larks (Grallinidae), bower-birds (Ptilonorhynchidae) and birds of paradise (Paradiseidae). The Australasian region is remarkable for its many specialized passerine families, and almost equal to the Neotropical region in differentiation from the rest of the world, as it is also in lowly mammals.

The Holarctic region has but three families of its own (apart from the loon order)—the auks (Alcidae in Laro-Limicolae) and the hedge-sparrows and tree-creepers (Prunellidae and Certhiidae, in Passeres). The New Zealand region has two families peculiar to it, the 'New Zealand wrens' and the wattle-birds (Xenicidae and Callaeidae in Passeres). The Oriental region has one, the fairy-bluebirds (Irenidae in Passeres). One family, the sheathbills (Chionididae in Laro-Limicolae) is restricted to the Antarctic and sub-Antarctic (not a full 'region').

Families not so far mentioned which are purely New World are New World vultures (Cathartidae in Falcones), turkeys (Meleagridae in Galli), the limpkin (Aramidae in Grues) and the vireos and tanagers and troupials, etc. (Vireonidae, Thraupidae and Icteridae in Passeres). Those which are purely Old World are bustards and button-quails (Otididae and Turnicidae in Grues), the crab-plover and the pratincole-coursers (Dromadidae and Glareolidae in Laro-Limicolae), sand-grouse (Pteroclidae in Columbae), frogmouths (Podargidae in Caprimulgi), rollers, bee-eaters, hoopoes and hornbills (Coraciidae, Meropidae, Upupidae and Bucerotidae in Coraciae) and in the Passeres the broadbills (Eurylaimidae), pittas (Pittidae), bulbuls (Pycnonotidae), wood-swallows (Artamidae) sunbirds (Nectariniidae), white-eyes (Zosteropidae), weaverbirds (Ploceidae), starlings (Sturnidae), Old World orioles (Oriolidae) and drongos (Dicruridae).

All the other bird families are found in both Old and New World, and the following are cosmopolitan, having at least one species breeding in every one of the six geographical regions, including New Zealand: Hawk-eagles and falcons (Falconidae and Accipitridae in Falcones), herons (Ardeidae in Gressores), swans, geese and ducks (Anatidae in Anseres), game-birds (Phasianidae in Galli), cuckoos (Cuculidae in Cuculi), rails (Rallidae in Grues), waders (Charadriidae in Laro-Limicolae), gulls, etc. (Laridae in Laro-Limicolae), pigeons (Columbidae in Columbae), kingfishers (Alcedinidae in Coraciae) and in the Passeres the great family Muscicapidae which includes flycatchers and thrushes; and the wagtail-pipits (Motacillidae).

New Zealand, the most impoverished region, lacks the following families which are otherwise cosmopolitan: osprey (Pandionidae in Falcones), ibises and storks (Threskiornithidae and Ciconiidae in Gressores), goatsuckers (Caprimulgidae in Caprimulgi), swifts (Apodidae in Macrochires), woodpeckers (Picidae in Pici); and in the Passeres swallows (Hirundinidae), finch-buntings (Fringillidae) and true crows (Corvidae).

It is obviously impossible in an introductory work such as this to trace the history of our knowledge of the avifauna of each region and sub-region of the world. Seventy years ago Newton (1884) did justice to the subject, as it then stood, only by dint of the utmost economy of words and judicious selection, and since his time the fantastic output of ornithology has made even a list of important summarizing syntheses impossible. But the following are the latest works which at least provide a list of the birds of the world's regions and sub-regions.

World: Sharpe and others (1874–98), Peters (1931).
New World: Cory, Hellmayr and Conover (1918–48).
Nearctic: Stone and others (1931).
Palearctic: Hartert and Steinbacher (1903–38).
Ethiopian: W. L. Sclater (1924–30).
Australasian and New Zealand: Mathews (1927–31).
New Zealand: Fleming and others (1953).

There is no special treatise on, or list of, the birds of the Oriental region, and no treatise on the birds of the Neotropical region apart from that of Cory, Hellmayr and Conover which also includes those of the Nearctic.

To illustrate the rise of geographical ornithology it is necessary, then, to take examples: and in the following pages I pay special (though not sole) attention to the birds of America. This is for several reasons, not least of which is the discreteness of the New World, and the fact that recent studies have been made of the two histories—the history of the American bird fauna (Mayr, 1946b), and the history of the American bird faunists (E. G. Allen, 1951).

In her interesting history of American ornithology before Audubon, Elsa G. Allen has traced the rise of early regional ornithology throughout the world, not merely in the Americas. She connects the rise of ornithology with the explorers.

> "During the period of frequent expeditions to Asia, to Africa, to the island world of the Pacific, and to the Americas there were vast new stores of knowledge fairly inundating the eager people at home. Practically every argosy that set sail carried artists and collectors whose duty it was to bring back drawings and specimens of birds, animals, and plants. Many of these poor creatures perished on the long voyages for lack of proper care, but such as survived were provided for by royal command or by the patrons of art and travel. It became customary for princes and titled persons to support these ventures for they realized their prestige was enhanced by their interest in science.
>
> People flocked to see these strange captive creatures and artists were engaged to paint them from life. Thus as the public became more familiar with living things, which formerly they had but vaguely visualized, books with improved illustrations came into demand. It was natural that birds with their multifarious colors should appeal as a subject to write about and to illustrate with colored engravings."

One of the earliest collections derived from voyages, museums and menageries was the *Historia Naturalis de Avibus*

(1650) of John Jonston [1603–1675], which is a compilation mainly from Aldrovandi except for its hand-coloured engravings of exotic birds. It still contained some mythical birds. For a century and more the obsession of most ornithographers was with the extraordinary spate of novelties and rarities that came their way from voyages of explorers, and thereafter voyages of trade and administration, all over the world. Museums and collections, private and public, began to accumulate. It was a long time before any sort of overall picture of the real avifauna of the great new colonies of the world began to emerge. It is of course impossible for an avifauna to be based on anything but a representative collection, which explains why Mark Catesby's *Natural History of Carolina* of 1731 (p. 65) is the first colonial fauna of any consequence. Even in the home countries of Europe there were no works which could be reasonably described as regional faunas before the seventeenth century.

Indeed, after the publication of Caspar Schwenckfeld's *Aviarium Silesiae* of 1603, no other comparable regional avifauna was published for over 100 years—not until Rzaczynski's book on Poland (1712), though there were what purported to be representative lists of birds earlier. For instance, there is a list, albeit incomplete even by contemporary knowledge, of British birds in Harrison's *Description of Britain*, which was first published at the end of Holinshed's *Chronicle* in the 1577 edition. Carew published a list of Cornish birds in 1602, Merret a list of British birds in 1667; and Sibbald's *Scotia Illustrata*, of 1684, almost qualifies ornithologically as a regional fauna. So, for that matter, do Plot's two county books, the *Natural History of Oxfordshire* in 1676 and of *Staffordshire* in 1686. In 1700 C. Leigh published a *Natural History of Lancashire, Cheshire and the Peak in Derbyshire*. But these lists and local natural history books, to which might well be added Morton's *Natural History of Northamptonshire*, 1712, were confined to small areas. When Gilbert White wrote in 1789, "Every kingdom, every province should have its own monographer", he was hopeful for the future, not appreciative of the past.

We have already noted the rise of our knowledge of British

birds since the pioneer work of Turner and Ray, and the rapidity with which, in the eighteenth and early nineteenth centuries, a practically complete identification of all but the irregular members of our avifauna was attained. We can contrast this with the development of our knowledge of what was, at least to begin with, colonial avifauna. Allen shows that it was Christopher Columbus himself (apart from the Norsemen who came from Greenland) who first made contact with the American avifauna from the Old World. Indeed, Columbus came across it before he discovered America, for it was because of the direction of the flight of the birds he encountered that he changed his course in 1492, and made his landfall in the Bahamas. Columbus appears to have been the first collector of American birds, for in April 1493 he showed several kinds of parrots and some bird skins which he had collected under the instructions of Queen Isabella of Spain.

Allen shows that among other early Americans who brought back live birds and specimens (usually parrots) and made notes was Oviedo y Valdés, who made six voyages to the Indies between 1514 and 1557, and who was the first to describe the American pelican, the ivory-billed woodpecker and a humming-bird. The first drawing on the spot of an American bird appears to be the picture of a turkey in South Carolina drawn by one Jaques le Moyne in 1564; and John White, on the second of Sir Walter Raleigh's expeditions to America under Sir Richard Grenville, made some fine drawings of American birds at the Roanoke colony, Virginia, in 1585, including sandhill-crane, flicker, pelican, noddy, flamingo, man-o'-war bird, booby and tropic-bird. Some of these he may have seen in the West Indies on the way to Virginia.

Many birds, including turkey-buzzards and condors, were mentioned by Joseph de Acosta, who emigrated to Peru in 1571, and in 1570–77 Francisco Hernandez described a number of birds from Mexico. The first native American to describe American natural history was Garcilazo de la Vega, son of an Inca princess and a Castilian father [1593–1615], who in his *Royal Commentaries of the Incas*[1], wrote of many different

[1]Works cited in this account of American ornithology are in the excellent bibliography of Elsa Allen if not in that of this book.

species with considerable verisimilitude; he was very "sensitive to bird song and was a good observer of bird habits", in Allen's words.

In North America there were several accounts of birds by the early fishermen of the Newfoundland Banks of the sixteenth century. Among the most important of these were Jaques Cartier, who made his first voyage in 1534, and Samuel de Champlain, who made his first in 1599. De Champlain writes with real familiarity of the birds of the Gulf of St. Lawrence, and indeed it is possible to use his work to establish what are clearly some subsequent range-changes of birds. Nicolas Denys, who went to the Gulf of St. Lawrence (Acadia) in 1632, made many important observations on birds which were largely forgotten until they were brought into prominence by E. G. Allen (1939) and F. H. Allen (1940). His contributions were certainly greater than those of any predecessor in the Americas. Other American voyagers who made useful notes and lists of birds were the Baron de Lahontan, who first came to Canada in 1683 and J. B. Labat, who became a colonial missionary in the West Indies in 1693.

There is quite a list of birds in the poems of William Wood's *New England's Prospect* of 1634 and many remarks on game and small birds in the work of Thomas Morton, *New English Canaan* of 1637. Many bird drawings were made by George Marcgrave, artist on the Dutch expedition to Brazil of 1660. In 1638–39 and 1664–73 John Josselyn made and published many notes on the birds of New England in two books, and John Clayton the elder, who was in Virginia in 1671, wrote many useful notes. But John Lawson, whose *History of Carolina* was published in 1714, two years after his death, produced a list exceeding anything published by earlier travellers, containing over a hundred recognizable species.

The next ornithologist to appear on the scene was the father of American ornithology, Mark Catesby [*c.* 1679–1749]. His *Natural History of Carolina, Florida and the Bahama Islands* (1731–43) contained 109 plates of birds, of which no less than seventy-one were subsequently used by Linnaeus for his descriptions. From the days of Catesby to the days of

Alexander Wilson [1766–1813] was a period distinguished largely by the work of European compilers on American collections sent or brought to them. Some of these resulted in fine descriptions and plates, particularly those of George Edwards [1694–1773], whose *Natural History of Uncommon Birds* (1743–51) and *Gleanings of Natural History* (1758–64) are fine collectors' pieces as well as works of ornithology.

By the time the tenth edition of Linnaeus's *Systema Naturae* was published in 1758, about 790 species of world birds could be mentioned, of which 260 were American; by 1889 the North American *Check-List* recorded 729 species, of which 202 were named by Linnaeus. By 1952 the A.O.U. list was about 793.

There are many American birds in Thomas Pennant's *Arctic Zoology* of 1784–87, and in John Latham's *Synopsis of Birds*, of 1781–1801, and his *General History of Birds*, 1821–28. All this time information about American birds continued to come from voyages, such as those of Captain Cook. Cook took with him naturalist-artists such as the Forsters, father and son, and W. W. Ellis. Explorers and fur traders kept up the supply of notes and specimens. Such were Humphrey Marten and Thomas Hutchins with the Hudson's Bay Company in the second half of the eighteenth century, and Captain George Cartwright of Labrador, who kept copious notes, including some on great auks, at the end of that century. Samuel Hearne, who travelled 1,300 miles on foot in the Canadian north-west in 1771–72, made many good observations. It is impossible to list the many other good bird-watchers of the early civilized years of North America, of whom Elsa Allen writes in her history. But I cannot refrain from mentioning that Thomas Jefferson himself published a list of birds in his *Notes on the State of Virginia* in 1787.

The first ornithologist to be born in North America was William Bartram [1729–1823], whose principal fame derives from his *Travels through North and South Carolina*, of 1791, and who wrote a catalogue of the birds of North America containing 215 species. Bartram was an able artist, as was John Abbott [1751–1840], who went to Virginia in 1773 and who made some hundreds of bird drawings which have never been published.

L. J. P. Vieillot [1748–1831], the famous French systematist, published a *Histoire naturelle des Oiseaux de l'Amérique septentrionale* in 1807–09, based on a stay in Hispaniola and San Domingo between 1790 and 1798. But it was Alexander Wilson [1766–1813] who would have become the greatest American ornithologist had he not died at forty-seven and had he not been overshadowed by the more artistic but less scientific ornithologist John James Laforest Audubon [1770–1851]. According to Gordon Wilson, quoted by Allen, Alexander Wilson made 320 figures of American birds, representing 262 species, of which thirty-nine were new to science and twenty-three others were proved by Wilson's description to be different from European species with which they had been identified.

Allen's sympathetic history and appreciation of Wilson's scientific work abundantly justifies her words.

> "It would be vain to attempt to remould the judgement of a century, but it should be pointed out that Wilson never took the position to which he was entitled both by priority and by certain scientific powers. Instead, Audubon, riding on the wave of bird interest that swept the late nineteenth century, has been lavished with praise, to the almost complete exclusion of a very able ornithologist.
>
> Audubon's much vaunted position is due in part to a fortuitous combination of circumstances which will not concern us here; but it is necessary that Wilson be cleared of the unpleasant penumbra of inferiority from which his reputation has suffered for over a century.
>
> It is true that his ambition was much greater than his ability to execute; but as a bird observer having the potential characteristics of a great ornithologist, he was pre-eminent at his time, exceeding all his predecessors, and even Audubon, in his ability to describe and interpret the living bird. There are many, no doubt, who will question this statement, but when Wilson and Audubon are compared as scientists and not as artists, Wilson's greater exactness, his patient method and his lucid and honest descriptions mark him unquestionably as the better ornithologist."

Alexander Wilson's *American Ornithology* was published in 1808–1814. From the ornithological point of view Wilson's great contribution was his establishing of the American avifauna as a special fauna. His was the first gathering together of information of the kind that could enable his successors to arrive at the now well-known principles of geographical distribution. Audubon mainly improved on Wilson by producing an artistic *tour de force*, although the 435 engravings in the *Birds of America* (1827–38) contained no less than 1,063 life-size figures of 489 supposedly distinct species, of which a few cannot be identified with existing known species. Audubon's book is, of course, the most famous bird book ever published. Its four double elephant folio volumes were originally published in eighty-seven parts, each with five aquatint plates coloured by hand and engraved in London (a few early plates in Edinburgh). It is unlikely that as many as 200 complete sets were bound up or that more than 300 impressions were taken of any one plate. It is an interesting commentary on values that the present worth of the published work of Audubon probably exceeds in cash £1,000,000 sterling; this is out of all proportion, needless to say, to its ornithological worth. The most recent collation of Audubonian is that of Sitwell, Buchanan and Fisher (1953). A great deal of Audubon's ornithological information derives from William MacGillivray, the great Scottish ornithologist, his mentor on many of his trips in Britain. Audubon was, of course, no mean naturalist, and conducted some well-designed experiments on the olfactory powers of birds of prey. But one cannot resist repeating that Wilson was a better bird man. One cannot even resist quoting Alfred Newton (1884) who said,

> "Audubon has been greatly extolled as an ornithological artist; but he was far too much addicted to representing his subjects in violent action and in postures that outrage nature, while his drawing is very frequently defective".

Perhaps we had better attribute Newton's remarks to the short patience of the academic man with anything that could be described as amateur work. But then Audubon's influence has not been purely ornithological. Every country seems to look for some hero round whom to build a natural history movement. In Britain we have been fortunate to find Gilbert White. If it had not been him, it would, I believe, have been Darwin, whose early adventures as a field naturalist have been so much overshadowed by the great generalization of his maturer years. In North America Audubon has been made the hero, and his name is used by a group of societies and by a movement towards the conservation, preservation, protection and better understanding of nature in its most general sense. The Audubon societies of America, and in particular the National Audubon Society, play a part and exert an influence in American politics which has no exact counterpart in Britain. The ambitious, obsessive and energetic artist, whose great bird book took the world by storm, could never have guessed some of the consequences of his existence.

It is good to think that Alexander Wilson has not been forgotten. His name is honoured by the *Wilson Bulletin*, one of a small group of scientific bird journals which occupies the place next to the *Auk*, and in American scientific records. It was in the *Wilson Bulletin* for 1946 that Ernst Mayr published his "History of the North American Bird Fauna". What a contrast there is between the synthesis of knowledge that Mayr found possible, and the state of knowledge in the days of Wilson and Audubon over a century before! It is true (and I have already made the point) that in the work of Wilson came the first real indication of the existence and nature of the American avifauna. The rest of the nineteenth century was spent in a scramble for collections and bird descriptions, which left few ornithologists time to think of anything but the most obvious kinds of systematics, and of the endowment of the museums of North America with magnificent and comprehensive collections—without which ornithology could not have taken the new directions of the twentieth century.

One of the results of the preoccupation of the nineteenth century with collection and description was the rather static view of the whole subject of geographical distribution. As Mayr points out,

> "Most of the geographical debates of the nineteenth century, after P. L. Sclater's pioneer generalization, consisted of arguments amounting occasionally to recriminations about the niceties of the status of the proposed regions and sub-regions."

The most telling *exposé* of what had become a stultification was that of the herpetologist E. R. Dunn (1922).

> "It is obvious that the zoogeographic realms are nothing but the great land-masses with lines drawn to correspond to the physiogeographic barriers. There is of course a great philosophical difference between such terms as holarctic fauna and holarctic region. In the first case we speak of zoological matters in terms of zoology, in the second of geographical matters in terms of zoology."

Since the turn of the century, indeed, the more enlightened of the systematists had been thinking of the fauna and not of the static regions. Gradually the idea dawned that the world was full of animal and plant communities each strikingly different from the other and each associated with a particular region of the earth's surface. But through history these communities have shifted and changed, mingled and separated, expanded and dwindled. The present zoogeographic regions of the world are simply the limits of the dominant communities. But it is the communities which identify the regions, and not the geographical regions the communities or faunas or floras.

Mayr's analysis cannot be quoted in detail. Amongst other things, it restores the Nearctic to something a little better than sub-regional status, or, as perhaps we had better say, sub-faunal status. He shows very neatly the effect of the sea barriers which broke Central America into two, so that there was no complete land connexion between North America and South America between the Lower Eocene about 60 million years ago and the Upper Pleistocene. He shows, too, that up to the mid-Tertiary a large part of the Nearctic was tropical, so that there was a North American radiation of tropical fauna, of which an important element survives. Mayr shows the existence of a pan-tropical element which includes the barbets, trogons and parrots. The trogons, which are most diversified now in Central America, probably originated in tropical North America (or the Oriental region),the parrots probably in the Old World, though they must have arrived also in the New World early, possibly before the Eocene separation of South and North America. The panboreal element is exemplified by the loon, phalarope and auk families, all of which are circumpolar but whose region of origin cannot otherwise be determined. In the whole of the Tertiary a connexion or near-connexion between Asia and North America across Bering's Strait provided a bridge for animals, and Mayr believes that a number of American bird families are an Old World element and invaded by this route. Some of these families have been so long in the New World that they have evolved into their own special New World sub-families (e.g. quails, cuckoos and gnat-catchers). Mayr identifies several separate invasions of America from the Old World by different groups of thrushes, which have proved very successful in the New World. Other groups of Old World origin with successful members in the New World are cranes, kingfishers, cardueline finches, tits, wagtails and pipits. Others have evidently colonized the New World only very recently and their few representatives have not greatly differentiated. Examples are barn-owls, larks, nuthatches, creepers, Old World warblers, goldcrests and shrikes. Of these, only larks and barn-owls have reached South America. Mayr believes that a number of important groups have originated in the North American continent, particularly during the tropical Tertiary period. He quotes Lönnberg in considering that the thrashers, mocking-birds, vireos, wood-warblers, waxwings, silky flycatchers, wrens and motmots are all North American, and adds the palm-chats, limpkins and New World vultures, also turkeys, grouse, dippers and buntings. Many of these families have invaded the Old World to a certain extent. Among the most successful of such invaders have been the waxwings, grouse, dippers and buntings, and one species of wren has made a rather complete conquest of the whole of the Palearctic region to remote oases of the Icelandic desert.

He next identifies a number of American bird families which are so rich in both South and North America that it is impossible to decide which was the region of origin. Among these, he thinks that the curassows and guans were originally northern, and the humming-birds, tyrant-flycatchers, tanagers and blackbird-troupials originally South American.

There is a long list of families more certainly of South American origin. They include besides the families mentioned on page 59, the honey-creepers, Coerebidae, and the cardinal subfamily Richmondeninae.

One of the most interesting points that emerges from Mayr's study is that not a single species of the originally South American fauna has crossed the Bering Strait bridge into the Old World, though quite a number of Old World birds have been able to invade South America. Another point is that the American avifauna has a really very different distribution from the American mammal or reptile fauna. And he shows that students of birds tend to find affinities between North and South America, whereas students of mammals and reptiles find greater affinities between North America and Eurasia. But as long as biologists realize that zoogeography is not ordinary geography, and that what they are studying are faunas and not regions, this interesting situation need not give rise to confusion.

CHAPTER VI

BIRD HISTORY

BIRDS originated from reptile stock some time at the beginning of the Secondary or Mesozoic era. The earliest known avian fossil, *Archaeopteryx* of the Jurassic, is almost sufficiently specialized for us to guess that the bird branch of the tree of life must have diverged from the trunk in the Triassic period, between 190 million and 150 million years ago, or perhaps even earlier, in the last period of the Paleozoic era, the Permian, the 30 million-year period when the first reptiles, freed from dependence on water, could exploit the vegetation-covered land. The evolution of feathered reptiles coincided with another reptilian experiment in flight, that of the pterodactyls, which, featherless and apparently scaleless and hairless, depended on a wing of skin stretched on a much-extended fifth finger and posteriorly to the hind limbs. The pterodactyls had two waves of evolution, the first mainly during the Jurassic (and coinciding with the first known fossil birds) when they were mostly small and long-tailed, and the second during the Cretaceous, the period of the most fantastic reptile radiation, which forms like *Pteranodon*, with a twenty-two-foot wing-spread, became the most exaggerated if not perhaps the heaviest of living things ever to take the air. *Pteranodon* must have been a sort of reptilian albatross, and convergently had gone through many of the avian evolutionary steps, with beak and hollow cantilevered bones. It had an extraordinary projection running posteriorly from its head, which some people think to have been a rudder. It could only have been a sea-pterodactyl, for with that wing-spread it could not possibly have carried enough muscular fuel for more than moderate journeys under power. It must have lived over fairly stormy seas where, like the albatross of today (which has half its wing-span), it travelled as a sail-plane. It must have lived on fish or marine animals. How it caught them; whether it could swim and if so, how;

how it folded its wings, or brought forth its young, we cannot guess.

Marine animals stand a better chance of being fossilized than most, since their bodies are more likely to be quickly covered with sediment. That may be one of the reasons why there are so many more pterodactyls known than Mesozoic birds. Another reason may well be that for that period in which the climate was decidedly warm the pterodactyls were more successful, the birds only really coming into their own when the sudden climatic change and its associated period of mountain-building put an end to the last period of the Mesozoic, the Cretaceous, and with it, at one fell swoop, the great and fantastic dinosaurs and pterosaurs. This is generally held to have occurred about 65 million years ago.

The earliest known birds, which as I have pointed out were certainly not the earliest birds, lived, then, in the Jurassic period, perhaps 130 million years ago. They had long tails consisting of up to twenty vertebrae and were probably only capable of flapping heavily from one branch to another. *Archaeopteryx* was found in 1861 by Haberlein of Pappenheim, named by Hermann von Meyer in 1861 and described in 1861 and 1862 by Andreas Wagner and in 1863 by Owen. Oddly enough this discovery did nothing to alter Wagner's famous anti-Darwinian views. The second bird, *Archaeornis*, was found in 1877 (Seeley, 1881). There are several suggestions as to the actual reptiles from which these birds are descended. Two of these are important. The first, associated with Nopcsa (1907), is that they were small, swift-running animals with long tails, whose forelimbs had progressively shortened and whose hind limbs had lengthened. If they existed they probably ran on their hind legs and did not jump like kangaroos. Nopcsa suggests that they flapped the forelimbs to help their progress and that the posterior margins of these limbs became broadened, not by a patagium (or stretched flap of skin) but first by scales and then by feathers derived from scales.

The other theory, perhaps more widely held, is that the ancestors of *Archaeopteryx* were themselves arboreal, and that broad-scaled and later feathered forelimbs were derived as an aid to gliding from branch to branch, like the wings of bats and the patagia of phalangers, flying squirrels, and the so-called flying lemur *Galeopithecus*. But it must be said that since in none of these flying mammals is the wing or patagium freed from the hind limb, there seems to be some reason for expecting the same in the wings of birds; Nopsca's cursorial theory, on the other hand, gets over this difficulty. Heilman (1926) suggests that Nopsca's bipedal reptiles themselves became arboreal; this 'compromise' might well be the solution.

Archaeopteryx was about the size of a crow. It had teeth in both jaws, a long feathered tail with about twenty-one vertebrae of which twelve were feathered, and three fingers separate from the wing, each provided with claws. (Claws still persist in the thumb and first finger of the ostrich, and in the young of the primitive South American hoatzin the thumb-claw exists and is used in climbing.)

The next period of the Mesozoic era, the Cretaceous, finds us at the stage where birds had become considerably advanced, having lost their horny tails, They still retained their teeth. Our knowledge of the important birds from this period is due to the researches of Marsh in America. The first of his discoveries, *Laornis* (1870), was a portion of a leg-bone of a bird the size of a swan; this bird, however, he could not place in any definite order. At the same time he described a genus of birds with probable affinities to the modern waders, *Palaeotringa*, and a possible rail, *Telmatornis*.

In 1872 Marsh (1872) announced his most important discovery, *Hesperornis regalis*, a large swimming bird. *Hesperornis* was three feet high, flightless, and with no keel on its breast-bone. It probably lived much as does a penguin, except that its breeding grounds were in the Northern Hemisphere and are now found quite high up on the Rocky Mountains, the heights of which, in those days, were quite low down, forming a chain of islands. Like modern divers, the Gaviae, to which it is probably closely related, it almost certainly could not stand upright on land. Its brain was very small. To *Hesperornis* Marsh added another bird, *Graculavus* (1872b), a relation probably of the modern cormorants.

Later in the same year Marsh (1872c) announced another important discovery, from the Upper Cretaceous shale of

Kansas, from which *Hesperornis* had come. This was *Ichthyornis*; it was quite unrelated to *Hesperornis* and may be quite a close cousin of the modern pelican-gannet-cormorant order, the Steganopodes. It was a foot high, toothed, strongly winged, and with a deep keel to its breast-bone. Another close relation of *Ichthyornis* he called *Aptornis* (1872d), and in 1893 he described a further *Hesperornis*-like bird, *Coniornis*. In 1873 he proposed the name Odontornithes (toothed birds) for the sub-class into which he put his discoveries; they are fully described in his monograph (1880).

Outside North America there have been few discoveries of fossil Cretaceous birds, though *Enaliornis*, a relation of *Hesperornis* and the divers, was discovered in Cambridgeshire (Seeley, 1866) and in Sweden a bird which may be an ancestor of the flamingo, *Scaniornis* (Dames, 1890).

Since these descriptions of Cretaceous birds were made in the nineteenth century, very little further evidence has come to light. The great adaptive radiation and success of the birds came early in the Tertiary or Cainozoic era, in the period of the Eocene, very swiftly. Birds at once became a dominant group of animals. In keeping with their rapid increase in species and numbers, there was of course an increase in the number of fossils. By the end of 1949, according to Wetmore (1951b), the greatest authority, there were probably 786 species of fossil birds known from the rocks. Yet, although the Mesozoic era was twice as long as the Cainozoic, only about a couple of dozen have come from the Mesozoic. Besides *Archaeopteryx* and *Archaeornis* in the Jurassic, five species are known of the *Hesperornis* order and ten of the *Ichthyornis* order. Another primitive flamingo, *Parascaniornis*, has been described, and the collection is completed by *Gallornis*, from the Lower Cretaceous, which is supposed to have been gooselike, *Elopteryx*, which is pretty certainly an ancestor of the gannets and cormorants, and *Enaliornis*, of the Lower Cretaceous. Two species of *Enaliornis* are now known, and the genus may be ancestral to *Hesperornis* (which belongs to the Upper Cretaceous) as well as to the modern divers and grebes. All the same, though these rare Cretaceous fossils undoubtedly foreshadow some primitive birds that survive today, it is necessary to quote

Wetmore: "the primitive beginnings of our living birds in Mesozoic time remain obscure"—to say the least of it! There is one other Cretaceous fossil which may be a bird, though it is known only from a single mandible. This is toothless and, as Wetmore points out, may possibly have belonged to a dinosaur, for some of the dinosaurs became 'bird-beaked'. If *Caenagnathus*, the name given it by R. N. Sternberg, is a bird, it should be placed with the Ratites, though, as Wetmore puts it, "on the basis of antiquity rather than on supposed close relationship".

The great strides in our knowledge of fossil birds from the Eocene onwards have been recorded in the important monograph of Lambrecht (1933) and in many papers by Alexander Wetmore and his pupil and colleague Hildegarde Howard, culminating in the useful summaries of Howard (1950) and of Wetmore (1951b). It is now possible to see the adaptive radiation of the birds in the Cainozoic era as a phenomenon of the same calibre as any of the other great revolutions in vertebrate evolution, like the conquest of the land by the amphibians, the freedom from dependence on water won by the reptiles, the fantastic radiation of the dinosaurs and flying reptiles, and the rise of the mammals. It must not be forgotten that throughout the whole of the Cainozoic or Tertiary era and in the Quaternary era (the last million years or so) the birds, except perhaps at the beginning of the Eocene, have probably always outnumbered the combined total of mammal and reptile species; and in two great areas of the world, almost though not quite distinct zoogeographic regions, the Arctic and the Antarctic, birds are the dominant form of vertebrate life.

In his presidential address to the International Ornithological Congress of 1950 Wetmore (1951b) said:

> "The very considerable increase in our knowledge of fossil birds that has been gained in the past seventeen years has served to support the idea that I have long held relative to the evolution of the living species of our modern world. I report here again my firm belief that birds as a class had their maximum abundance in Tertiary ages, and that our living kinds had their evolution both as genera and as

species in the Miocene and Pliocene periods. The space of Quaternary time that followed has been one of rigor, pressure and extermination, beginning with the Pleistocene, when sheets of ice covered large areas in the Northern Hemisphere where formerly birds had found congenial environments for increase and diversification, and culminating in our present age when Man as the dominant creature has overrun the Earth, converting it to his own uses, not only destroying by hand many of the birds and other creatures that formerly flourished, but also usurping and changing the land so that great numbers of birds have been destroyed because of loss of their proper habitats."

It seems unquestionable that birds have been reduced by the Pleistocene glaciation, though it would seem that mammals were reduced in species and populations by the same cause, and probably more. Now that the animal community has emerged from the Ice Ages into the Recent period characterized by man's domination of the earth's surface and all that grows on it, we must agree that man has distinguished himself as the most important single agent for the destruction of bird species, directly or indirectly. In historical times some hundreds of species have become extinct in circumstances that make it pretty clear that man was responsible. However, this process is not necessarily a continuous and inevitable one. Moreover (see p. 120), species continue to become extinct through no fault of man's and to emerge with no assistance from him. It is also now probable that with newly found and strong emotional appreciation of birds and scientific interest in them, man may be able to arrest some of the processes of extinction for which he has been responsible: indeed he has already begun to do so in some cases.

This digression has led us some distance from our development of the history of birds in the rocks of the Cainozoic or Tertiary era. In the Eocene period, which lasted from 60 million or 65 million to 40 million or 45 million years ago, the evidence of the rocks shows a rapid transformation and radiation of birds, with great changes in the 20 million years. In the Cretaceous all the few birds we know of were toothed, though some, like the rather numerous *Ichthyornis* species, were otherwise birdlike, tailless and with keeled breast-bones. (*Hesperornis* did not have a keeled breast-bone, but it was flightless.) Indeed, the only Cretaceous bird that had no teeth was *Caenagnathus*, if it was a bird.

None of the numerous species known from the Eocene has teeth. These species belong to fifteen orders, all but two of which, the Diatrymae and the Aepyornithes, have members living today, and it is doubtful whether the order Aepyornithes became extinct until historical times. Many of the orders of the Eocene are represented by several families, and the families often by several genera. Only one or two of these genera is the same as that of any living bird, though since the status of a genus is anyhow so much a matter of opinion, it would not stretch unduly the imagination of a competent systematist if some of the Eocene game-birds and cranes were to be put in the genera *Coturnix* or *Grus*. It is noteworthy that the Eocene birds include several members of modern families of the Passeres, considered to be the 'highest' order of birds. Beecher (1953) points out that, on distributional evidence only, the Passeres, and in particular their most primitive family, the Sylviidae or Old World warblers, may date back to the Cretaceous. Flowering plants evolved in the upper Cretaceous, and so these birds, probably the only generalized insectivorous birds of that time, could have a quick success and wide adaptive radiation, and become by far the biggest order.

Generally it is the late Eocene forms of birds which belong to modern families or resemble modern genera. The early Eocene birds, of which there are not very many, are obviously more primitive and less clearly differentiated, some forms, like *Colymboides*, being intermediate between and presumably ancestral to two modern orders (such as, in this case, the loons, Gaviae and grebes, Podicipedes).

Although the Tubinares (tubenosed birds) is obviously an ancestral order, there is no Eocene example, unless *Gigantornis eaglesomei*, from the Nigerian deposits, is, as Lambrecht suggests, between the Tubinares and the other primitive sea-bird group, the Steganopodes. *Gigantornis* is known only from its breast-bone, but this is so vast that its wing-span may well have been about twenty feet. This is bigger than has been known in any flying bird, ancient or modern, and closely rivals that of *Pteranodon* (p. 73), the most exaggerated winged animal. Like *Pteranodon*, and like the modern albatrosses, of which it may have been a relative, *Gigantornis* must have been a sea-going sail-plane. There are several examples of true Steganopodes in the Eocene, from *Argillornis* of Owen (1878) to *Eostega*, intermediate between modern gannets and cormorants and, as suggested by Howard, on the line of descent from the Cretaceous *Elopteryx*. There was a *Protopelecanus* not far from modern pelicans and a *Protoplotus* not far from modern darters, and a *Prophaëthon* giving evidence of the divergence of the tropic-birds from the primitive cormorant-gannet stem. *Odontopteryx* (Owen, 1873) is placed in a family of its own among the Steganopodes. Its jawbones were strongly serrated, so that it had a toothed appearance, though its teeth were 'false' and secondarily developed in evolution. The divergence in the Eocene is equally well shown by the members of many other orders, of which a number can be assigned to present day sub-families. Thus among the Gressores, which include the modern herons, storks and ibises, we find the herons *Proherodias* and *Eoceornis*, and the bittern *Botauroides*. There was an ancestral flamingo, *Elornis*, belonging to the order Phoenicopteri.

In the Upper Eocene there are anseriform birds, putatively descended from *Gallornis* of the Cretaceous; *Romainvillia* and *Eonessa*, the former an ancestor of geese and pochards perhaps, the latter perhaps related to the modern ruddy ducks. The exact relationship of the earliest probable bird of prey of the Eocene, *Lithornis*, is obscure, but by the Upper Eocene we find that the order has already split into its present-day families —the Cathartidae, or New World vultures, represented by *Eocathartes* and *Plesiocathartes*; the Accipitridae, hawks and Old World vultures, represented by *Palaeohierax*, quite close to the present Old World vultures, and the eagle-like *Aquilavus*. The alleged presence of the genus *Aquila* (eagle) in the Eocene is unproved; at least one fossil supposed to belong to this genus was actually the claw of an edentate mammal! And the identification of the genus *Falco* is doubtful (Stone and others, 1931).

The interesting *Neocathartes* of this period has been placed by Wetmore in a family of its own.

From the middle Eocene we find quite a number of game-birds, the earliest being *Gallinuloides*, belonging to a now extinct family of the order Galli, the Gallinuloidae, *Filholornis* is an ancestral form which may be between the present family Cracidae (curassows) and the hoatzin (Opisthocomidae). *Palaeophasianus* appears to be a true tetraonid (the grouse-partridge family), and a quail, possibly of the modern genus *Coturnix*, belongs to the Phasianidae or pheasant family. Among the Grues, the modern order which includes the cranes, rails, cariamas, bustards, etc., two extinct families are represented by *Eogrus* of the Upper and *Geranoides* of the Lower Eocene. True members of the Gruidae or modern crane family are quite numerous in the Middle Eocene. There is a number of species of the genus *Aletornis*, and other recognized genera are *Paragrus* and *Palaeogrus*. Shufeldt (1915) claimed that a crane of the Eocene actually belonged to the modern genus *Grus*, but this is doubtful. Also in the Eocene we find *Telecrex*, which may be ancestral to the rails (Rallidae) and limpkins (Aramidae), and three genera appear to be unequivocal Rallidae —*Telmatornis*, *Palaeorallus* and *Eocrex*. There is also a genus from the middle Eocene, *Palaeotis*, which has been placed in the Otididae or bustards. Among the gull-wader order, Laro-Limicolae, radiation in the Eocene reached many present families. *Presbyornis* is allotted by Wetmore to an extinct family, but *Rhynchaeites* probably belonged to the modern family Rostratulidae (painted snipe), *Palaeotringa* to the sandpiper family Scolopacidae, and *Halcyornis* possibly to the gulls. The auks, Alcidae, are also represented in the Eocene by a now extinct subfamily, the Nautilornithidae, represented by three species belonging to two genera, *Nautilornis* and *Hydrotherikornis*. There are Eocene owls, even; in the Lower Eocene several species of *Protostrix* belonging to the extinct family Protostrigidae which Wetmore places nearer the Strigidae than the Tytonidae; in the Upper Eocene a strigid owl certainly belongs to the modern genus *Bubo*. There is a mid-Eocene hornbill (family Bucerotidae, order Coraciae) known as *Geiseloceros*. The highest order, the Passeres, is represented in the Eocene by a genus *Osteornis* which appears to be in the shrike family Laniidae, and by *Palaegithalus*, a little bird which is somewhere between the Old World warblers, Sylviidae, and the tits, Paridae.

Our survey of the birds of the Eocene is now complete, save for the known Ratites and one anomalous group. Of the ratites there is evidence of a bird which appears to belong to the same order, Struthiones, as the ostriches, though it has been placed in the extinct family Eleutherornithidae. *Eleutherornis* was described from a fairly well-preserved pelvis, and Wetmore says it "offers some peculiarities but seems to belong in an ancestral group that has produced the living ostriches". Another flightless bird of the Eocene, *Psammornis*, probably belongs to the Aepyornithes. For some time a group of biggish flightless birds of the Eocene were thought perhaps to be Ratites. Thus points of resemblance between *Dasornis* and the ostrich were pointed out by Owen. However, it now seems that *Dasornis* and its contemporaries *Remiornis* and *Gastornis* should belong to the Gastornithidae, and with another running bird *Diatryma* (Diatrymidae) should be placed in the extinct order Diatrymae which may lie between the cranes and the gull-waders. *Gastornis* was larger than an ostrich and *Diatryma* stood seven feet high.

It is not necessary to describe the avifauna of subsequent periods in such detail, for we have proved our point that the primary adaptive radiation of the birds was in the Eocene. But the following events in the Oligocene (45–33 million years ago) are worthy of note. There was still a form *Colymboides* intermediate in character between grebes (order Podicipedes) and loons (Gaviae), but true grebes (*Colymbus*) and divers (*Gaviella*) were also found. A doubtful form, *Hydrornis*, might be between the Anseres and the Tubinares; but, in the latter, *Manu* appears to be an albatross of the modern family Diomedeidae, and the modern genus *Puffinus* (shearwaters, Procellariidae) is found. So is the modern genus *Morus*, among the Steganopodes, the genus of the gannet, and so is *Phalacrocorax*, the cormorant. Wetmore (1951, p. 2) places two anomalous genera, hitherto thought to be penguins, in a special family of the Steganopodes, Pelagornithidae. Among the flamingoes, *Elornis* is still found, but also 'improved models', including the modern genus *Phoenicopterus*. Among the Gressores we find our first true storks (*Propelargus*) and ibises (*Ibidopsis*) and a now extinct family embracing the genus *Agnopterus*. Ducks and geese include now extinct genera, but *Aythya*, the genus of the pochard, and *Anas*, the genus of the mallard, are now with us. Among the cathartid Falcones there are several forms, and among the accipitrid Falcones one might almost say many, with recognizable hawks, eagles, and buzzards of the modern genus *Buteo*. In the Galli we find the modern pheasant genus *Phasianus*. Among the Grues are cranes, limpkins and rails; the extinct family of *Eogrus* continues, and in South America this order evolved a group of extraordinary families, which were once generally classed as the family Phororhacidae, but is erected into a suborder and divided into five families, which Wetmore (1951, p. 17) places next to the allied suborder Cariamae, which contains three families, two of which are also fossil, the third of which is the modern cariamas. This interesting group of fantastic specialized forms was investigated in the Santa Cruz and other Tertiary formations of Patagonia, particularly by F. Ameghino (1891), also by Moreno and Mercerat (1891). The best-known genus, *Phororhacos*, had a very large head and a small body, looking like a fantastic caricature of a cariama or secretary-bird. One of its species had a head two feet long with an extraordinary deep beak, hooked at the end, like a gigantic eagle; it was between seven and twelve feet high, could not fly, and, like the modern cariamas, pursued its prey by swift running. Another big form, *Andrewsornis*, had a mandible nearly sixteen inches long. These queer birds persisted into the Pliocene, when they died out.

Among the Oligocene Laro-Limicolae we find several modern genera—e.g. *Tringa* ('shanks'), *Limosa* ('godwits'), *Calidris* (knots) and *Charadrius* (ringed plovers); and we find two among the owls, *Strix* and *Bubo*. As in the Eocene, we find a hornbill (*Cryptornis*); there is a possible woodpecker (Pici order), and shrikes and wagtails (Motacillidae) among the Passeres. In the lower Oligocene there was an *Aepyornis*-ancestor, *Eremopezus*.

It can be seen that the Oligocene was a period of orderly development, with perhaps only one big experiment—the South American running gruine suborder. Much the same goes for the Miocene; the period of Alpine mountain-building, which shook large parts of the world at the end of the Oligocene and the beginning of the Miocene, appears to have had singularly little effect on the course of avian evolution. In the Miocene we find the first certain penguins; for Simpson (1946) has shown that supposed Eocene and Oligocene records, such as the seven-foot New Zealand *Palaeeudyptes* of T. H. Huxley and Hector, are of later periods or not of penguins.

Modern genera identifiable in the Miocene include *Gavia* (Gaviae), *Fulmarus*, *Puffinus*, *Diomedea* (Tubinares), *Morus*, *Sula*, *Phalacrocorax* (Steganopodes), *Ardea*, *Leptoptilos* (Gressores), *Cygnus*, *Anser*, *Aythya* (Anseres), *Haliaeëtus*, *Falco*, *Urubitinga* (Falcones), *Ortalis*, *Cyrtonyx*, *Phasianus*, *Tympanuchus* (Galli), *Otis* (Grues), *Charadrius*, *Limosa*, *Numenius*, *Himantopus*, *Larus*, *Uria*, *Cerorhinca* (Laro-Limicolae), *Strix* (Striges), *Conuropsis*, *Psittacus* (Psittaci) and *Corvus* and *Fringilla* (Passeres). For the first time we find sand-grouse and pigeons (Pterocletidae and Columbidae, Columbae), turacos (Musophagidae, Cuculi), swifts (Apodidae, Macrochires) and trogons (Trogones) in the rocks. In the Miocene also are the first-known New Zealand moas (Dinornithes).

In the Pliocene period (13–1 million years ago) we find as might be expected, more modern genera, and one fossil is osteologically indistinguishable from the modern crane species *Grus canadensis*, the little brown or sandhill-crane of North America. We meet the first ostriches, pelicans of the genus *Pelecanus*, black geese (*Branta*), king-vultures (*Sarcoramphus*), and the South American Phororhacids have their last fling, as do the Hermosiornithids, another gruine group closer to the cariamas. Among the auks, *Brachyramphus*, a modern genus, is found, as is a curious large puffin, *Pleiolunda*, and a flightless form, *Mancalla*, which, as Miller and Howard (1948) show, had progressed much farther towards the development of a truly penguin-like swimming wing than had the recently extinct great auk, *Pinguinus impennis*.

The period of the Ice Ages, the Pleistocene, which lasted

certainly for considerably over half a million years, and probably for a million, is full of avian fossils. Of the 176 Pleistocene species listed by Wetmore in the 1931 *Check-List of North American Birds*, no less than 130, or three-quarters, can be identified with surviving species. Indeed, the avifauna has probably changed less since the Pleistocene than has the mammalian fauna, at least in the Holarctic region.

Howard finds the representation of modern forms in the Pleistocene so great that "it leads to the conclusion that, at least by Upper Pleistocene time, all present-day types of birds were essentially established;" and she goes on to point out that the number of the species of the Pleistocene which are now extinct have been over-emphasized in studying Pleistocene deposits, because of their peculiarities. In Upper Pleistocene deposits such as those of Rancho la Brea in California or Fossil Lake in Oregon, the modern forms exceed the extinct by about five to one (Howard, 1947). However, some of the now extinct birds of the Pleistocene are of rather special interest. They include a *Leptoptilos* stork from Java which may have stood six feet high or more, and a *Stercorarius* (skua) from Oregon which is probably ancestral to the four known living forms. Wetmore has described a huge eagle, *Titanohierax*, and a huge owl, *Tyto pollens*, from the Bahama Islands. These probably lived on the large extinct rodent, *Geocapromys*, whose bones were mingled with those of the birds; and he has discovered evidence of an even larger owl that inhabited Cuba in the Pleistocene.

Although the moas of New Zealand go back to the Miocene, their most important radiation took place during the Pleistocene. The first discovery of bones of the moa was made near East Cape in the North Island of New Zealand, and such bones were described in 1838 by Owen (1839, 1842) and assigned by him (1843) to *Dinornis novaezealandii*. When this great anatomist was first shown a moa bone in 1839, he had seen no bird larger than an ostrich, and at first refused to believe either that a bird could have grown such a bone or that the bird could have lived in New Zealand. Upwards of half a dozen species of *Dinornis* have since been described, the largest of them having stood probably twelve feet high; and six other genera have been found, some as big as *Dinornis* but others only three or four feet high. They were adapted probably to quite different types of habitat. These genera have been described by von Haast (1886), Reichenbach (1849–50), Lydekker (1891) and Forbes (1892); for a good list of moas (and Aepyornises) see Rothschild (1907) and Buick (1931); there is also a useful description in Archey (1941). Hutton (1892) in his long paper on the moas produces sufficient proof that they were alive at the time of the Maori colonization. He says "We must conclude that in the North Island the Moa was exterminated by the Maoris not very long after their arrival in New Zealand—that is, not less than four or five hundred years ago".

Most modern authorities, including Archey (1941) accept the theory that the moas were exterminated by the earliest men to reach New Zealand, so that they had probably disappeared before the ancestors of the present Maori tribes arrived in A.D. 1350. However, Buick's interesting discussion suggests, though it does not prove, that the moa may have survived much longer in the South Island than in the North, probably until the sixteenth and possibly even until the nineteenth century.

The most spectacular deposit of moa bones ever discovered was found at Pyramid Valley, in North Canterbury, in 1937, and is described by Duff (1949). It had been over fifty years since a moa swamp of this kind had been found. By 1941 no less than fifty more or less complete skeletons had been excavated of the species *Dinornis maximus*, the greatest of the moas, *Pachyornis elephantopus*, *Euryapteryx gravis* and *Emeus crassus*. It was the Pyramid Valley collection which enabled Archey to revise the moas. Eventually, by 1949, 140 specimens had been excavated. *Dinornis*, of which there were forty-four, stood from ten to twelve feet high, *Pachyornis*, of which there were fourteen, from five and a half to seven feet, *Euryapteryx*, of which there were twelve, four and a half to five and a half feet, *Emeus*, of which there were seventy, the same height. *Pachyornis* and *Euryapteryx* had relatively short and very massive legs. Other interesting birds were found in the swamp: *Aptornis* was a large flightless rail, two to three feet, related to the modern weka. There was also a flightless goose, *Cnemi-*

ornis, and a great eagle, *Harpagornis*, with a six-inch head and probably a seven and a half foot wing-span, which probably fed on the moas.

Amadon (1947) has estimated the weight of the largest moa, *Dinornis maximus*, at 236 kilogrammes, or about 520 lb. Although this may have been the tallest bird that ever lived, it was certainly not the heaviest; this distinction is reserved for the largest elephant-bird of Madagascar, *Aepyornis maximus*, which Amadon puts at 438 kilogrammes or about 965 lb., not quite half a ton. The first scientific description of an enormous ratite bird from Madagascar is that of Isidore Geoffroy Saint-Hilaire (1851), but there is previous evidence from a book by E. de Flacourt (1661). De Flacourt says, "*Vouron patra*, C'est un grand Oyseau qui hante les *Ampatres*, et fait des œufs comme l'Autruche, c'est une espèce d'Autruche, ceux desdits lieux ne le peuuent prendre, il cherche les lieux les plus déserts". Our next clue (Strickland, 1849) comes from the diary of a Mr. John Joliffe, surgeon of *H.M.S. Geyser*, which was cruising off St. Augustine's bay, Madagascar, in October 1848, with a French merchant of Bourbon, M. Dumarèle, on board.

> "M. Dumarèle casually mentioned that some time previously, when in command of his own vessel trading along the coasts of Madagascar, he saw at Port Liven, on the north-west end of the island, the shell of an enormous egg, the production of an unknown bird inhabiting the wilds of the country, which held the incredible quantity of *thirteen wine quart bottles of fluid*!!! he having himself carefully measured the quantity. It was of the colour and appearance of an ostrich egg, and the substance of the shell was about the thickness of a Spanish dollar, and very hard in texture. It was brought on board by the natives (the race of 'Sakalavas') to be filled with rum, having a tolerably large hole at one end, through which the contents of the egg had been extracted, and which served as the mouth of the vessel. M. Dumarèle offered to purchase the egg from the natives, but they declined selling it, stating that it belonged to their chief, and that they could not dispose of it without his permission. The natives said the egg was found in the jungle,

and observed that such eggs were *very very rarely* met with, and that the bird which produces them is still more rarely seen."

On 27th January, 1851, Isidore Geoffroy Saint-Hilaire (1851) showed the French Académie des Sciences two eggs which had been discovered in 1850, by a ship's captain, M. Abadie, on the south-west coast of Madagascar. He said, "M. Abadie saw one day a gigantic egg in the hands of a 'Malgache', which the natives had perforated at one end, which they were using for various domestic purposes". Abadie offered rewards for more eggs, and succeeded in getting another from recently deposited rubbish in the bed of a stream. A third egg and some large sub-fossil bones were later found.

Saint-Hilaire was struck at once by the resemblance of the two eggs that arrived whole and a metartarsal bone that arrived with them to those of *Dinornis*, the moa. He gave the new bird the name *Aepyornis maximus*. (See also Milne-Edwards and Grandidier, 1866–73.)

This was, as a matter of fact, the stoutest of all the elephant-birds, though another species, discovered somewhat later, may have been taller. There are two smaller genera, one of which had bones very much more slender than the elephantine type shown by *Aepyornis*, and was probably more like the modern cassowaries (Rothschild, 1907). Marco Polo in the thirteenth century recorded that the roc, thought to be a legendary bird, came from Madagascar; it seems that it was not so legendary.

CHAPTER VII

BIRD SPECIATION

"A SPECIES," writes Ernst Mayr (1951) "is an aggregate of interbreeding natural populations which are not only reproductively isolated from other such aggregates but also ecologically specialized sufficiently so as not to compete with other such species." This is the latest and most succinct definition by the greatest authority on avian speciation[1].

It is interesting that this definition of Mayr's makes no mention whatever of morphological characters. It is the product of the 'new' systematics, in which the approach is biological and in which evidence is brought to bear not only from morphology but also from the study of genetics, distribution, behaviour, parasites, physiology and other subjects. The old systematics had as its definition of species something like this: "A species is a group of individuals or populations with the same or similar morphological characters." As Mayr (1942) says, "The trouble with such a definition is that it does not delimit true species from subspecies below or genera above". The reality of species is something which is very clear to the experienced museum systematist—and often clear even to the experienced bird-watcher who is not interested in museum work and does not handle skins. Yet it is extraordinarily difficult to define a species. Mayr's definition, which we quoted at the beginning of the chapter, is a good one and the product of a lifetime's experience and study. But it cannot be applied simply and quickly to all situations that are encountered, far less a box of skins in a museum. From the practical point of view we are still largely with the pragmatic views of Tate Regan (1926) who in a now classic passage suggested that "a species

[1]In a recent paper, seen by the writer when this book was in proof, Cain (1953) attempts to define a species without the use of the ideas of 'geographical' and 'ecological' distribution, which he believes can sometimes inextricably overlap. His ingenious arguments are food for the thoughtful evolutionist rather than a guide to the practical systematist, I suggest.

is a community, or a number of related communities, whose distinctive morphological characters are, in the opinion of a competent systematist, sufficiently definite to entitle it, or them, to a specific name."

This definition was not intended to be facetious, and will continue, I believe, to be taken seriously for many years. And rightly so, if the unanimity of those present systematists whose competence is acknowledged is any guide. The similarity of the opinions of these workers on 'difficult' groups is remarkable. The student is recommended to compare the number of species considered to belong to the non-passerine bird families by the late J. L. Peters (1931 onwards), and by Mayr and Amadon (1951) in their latest census. Neither the distinguished committee of the American Ornithologists' Union in its *Check-List of North American Birds* (1931) nor the distinguished subcommittee of the British Ornithologists' Union in their *Check-List of the Birds of Great Britain and Ireland* (1952) appear to have the slightest difficulty in agreeing upon which members of the lists were entitled to full specific status. Moreover, allowing for the two species admitted since the publication of the late H. F. Witherby's *Check-List of British Birds* (1941), the 1952 British list, as regards species, coincides with its distinguished predecessor precisely. Such unanimity would probably not be possible in any other branch of the animal kingdom, and be quite impossible among botanists. The only questions slightly in doubt about any birds on the British list seem to be: whether the pink-footed and bean-geese should be regarded as subspecies of a rather deeply differentiated species, or allopatric species belonging to a homogeneous superspecies; and whether the hooded and carrion-crows (p. 104) should be considered conspecific.

It is here necessary to introduce two valuable terms. However much one may deprecate the building up of a private language in which to discuss biological subjects, there is no doubt that the words *sympatric* and *allopatric*, which were coined respectively by Poulton (1903) and Mayr (1942), have abundantly earned their keep. Two forms or species are sympatric if their areas of distribution overlap or coincide. Two forms or species are allopatric if they do not occur in the same area. Generally these terms are applied to two breeding ranges of birds. For instance, many species and races whose breeding areas are separate and distinct from each other, overlap in their winter range and may indeed feed or flock together. So while the practice may be less strict than some would have, the word allopatric will be used in this book for those birds whose breeding ranges are discrete. We will discover that the differences between allopatric species are much more likely to be subtle and slight than those between sympatric species, which are nearly always very distinct and which give little trouble to the systematists. It is the allopatric species that teach us most, incidentally, about the process of speciation itself; and the fact that even the greatest systematists in the world cannot always distinguish between allopatric species and well-marked subspecies, simply means that evolution is a process and that the birds concerned have been surprised by ornithologists in the very act of crossing the invisible threshold between race and species.

This leads us to an important point: to the fact so abundantly maintained by Mayr for many years and finally proved by him in 1942, that, among the birds at least, species are derived from geographical subspecies. The three possible theories of the origin of species are that the fundamental differentiation is geographical, that it is ecological, or that it is genetic. Mayr has carried the majority of present workers with him in his unequivocal stand for the geographical origin of species. He maintains that it is geography and geography alone, not even a combination of geography and genetics, or geography and ecology. The genetic and ecological differentiation that is associated with speciation and which he analyses with such great care can be always proved, in his view, to be subsequent to and a consequence of the initial geographic isolation of the potential species. Some cases of speciation, it is true, appear to have been the result of a basic ecological differentiation; but in all such cases the differentiation can be shown also to have been clear-cut in space, although on a narrow margin.

It is very clear that the process of speciation is not uniform. It is slow in certain orders of birds, rapid in others; slow among birds with certain habits, rapid among those with others; slow in certain parts of the world, rapid in others.

Birds, like all animals, vary. It is a commonplace that the size and colour, and other characters of the form of an individual bird, can be modified by its food and other environmental factors subsequent to its conception. But it is with inheritable variation that we are concerned when we examine speciation. For it is the inherent genetic variability of animals that forms the stuff of evolution.

Geneticists have not so far studied the genetical variation of wild populations of birds extensively. Mostly they have worked on domestic species—fowl, pigeon, budgerigar, canary and others. The populations of these birds are nearly always subject to unnatural selection though of course to natural selection as well. The range of variability in size among domestic birds does not appear to be very great, but in plumage it is as extraordinary and as well marked as in any other class of the animal kingdom. The Mendelian inheritance of the plumage characters of canaries, budgerigars and some love-birds is well known. Many types of variation are common to all or most birds; such as albinism, involving a total lack of pigment, even from the eyes, melanism, which is excess of black pigment, xanthrochroism, where birds keep their natural yellow pigment but lose all or part of their black pigment, and erythrism, in which red predominates. Many species of the parrot family, normally green in the wild, throw well-marked and well-known blue sports in captivity. The genetics of the interrelation of many plumage and pigment characters of domestic birds is quite well-known, this knowledge now extending through the world of the professional cage-bird fancier.

In wild populations of birds, striking variations of plumage are not the rule; but in some wild species well-marked variants sometimes occur in such numbers as cannot be accounted for by the ordinary chances of mutation. The occurrence of two or more distinct phases or types in a population is known as polymorphism. Among several species of herons, for instance, albinistic, melanistic and erythristic variants are found in wild populations far more often than could be expected by chance.

Many of the tubenosed birds are dimorphic or polymorphic, having dark and light phases and sometimes intermediates between them. Among these are the giant fulmar, or stinker, *Macronectes giganteus*, the northern fulmar, *Fulmarus glacialis*, some of the gadfly-petrels of the genus *Pterodroma*, notably the capped petrel *P. hasitata* (whose dark form, which once bred in Jamaica, and has been known as *P. caribbaea*, is probably now extinct), *P. neglecta* the Kermadec petrel and *P. arminjoniana* of south Trinidad with its Pacific representative *P. a. heraldica*. The Kermadec and Trinidad petrels, which are also closely related, have dark- and light-breasted forms distributed apparently evenly throughout their range, but among the herald-petrels, the Pacific race of the Trinidad petrels, the dark forms make up ninety-nine per cent of the population of Henderson Island in the Tuamotus, less than ten per cent on Ducie Island, about 220 miles away, about one per cent in the Marquesas, and are apparently absent elsewhere.

At least seven species in the heron-bittern family show some form of dimorphism. It is not common among the ducks and geese, but the blue snow-goose and snow-goose have been shown to be dimorphic phases of the same species. Polymorphism is also known amongst the birds of prey. In Australia, for instance, there are two phases of the sparrow-hawk *Accipiter novaehollandiae*. The gyr-falcons of Greenland and North America, which are now usually lumped together as *Falco rusticolus obsoletus*, are trimorphic, the three principal phases being connected by intermediates; they are well described by Schiøler (1931) and Salomonsen (1951); they are dark, grey and white. The skua family is productive of phases: the arctic skua, *Stercorarius parasiticus*, has a dark phase almost uniformly brown and a light phase with buff and yellow underparts; the pomarine skua, *S. pomarinus*, likewise has dark and light phases; among the long-tailed skuas, *S. longicaudus*, the dark phase is exceptionally rare. The common guillemot, *Uria aalge*, has an interesting phase with a white 'spectacle' or bridle round the eye. In parts of the arctic range of the black guillemot, *Cepphus grylle*, there is a rare black-winged phase, and what is either an allopatric species or a well-marked subspecies in the Sea of Okhotsk, *Cepphus carbo*, has a wholly black-winged population.

As we search through many of the other orders and families of birds, we find several polymorphic species—such as the fruit-pigeon, *Macropygia mackinlayi*, the long-tailed New Guinea lory, *Charmosyna papou*, which has a dark phase. Among the owls, e.g. the short-eared owl and the brown owl, are found grey and erythristic phases; a kingfisher, *Halcyon saurophaga*, is polymorphic; so is a barbet, *Lybius torquatus*, dimorphic with a red or a white head. Among the flycatchers dimorphism may be found amongst the paradise-flycatchers, *Terpsiphone*, in the monarch, *Monarcha castaneoventris* (which sometimes has a black belly); the pied fantail *Rhipidura flabellifera* has a melanistic phase; and polymorphism is found among thrushes of the genus *Copsychus*, and *Oenanthe* (wheatears). Some dippers (*Cinclus*) are dimorphic, and the yellow wagtail group, *Motacilla flava*, is undoubtedly polymorphic, as a result of which most accounts of its systematics have become confused. There is dimorphism among the shrikes of the genus *Chlorophoneus*, in the banana-quit, *Coereba flaveola*, and the butcher-crow *Cracticus quoyi*.

One of the most interesting cases of dimorphism is that of the pretty Australian Gouldian finch *Poëphila gouldiae*, which exists in two common phases, one in which the facial mask is completely black like the rest of the head, and another in which this mask is red. H. N. Southern (1945) has shown that it is probable that the colour of the face-mask is caused by a single sex-linked gene, the factor for red being dominant to that for black.

It must be stressed that this well-known polymorphism in certain birds constitutes a different kind of variation from the occasional appearance of mutant sports. The British bird from which mutant sports are most frequently recorded is perhaps the blackbird, *Turdus merula*, examples from which are often partly white and occasionally wholly albino. The frequency with which they are reported compared with the freaks of other species is simply a consequence of the abundance of the blackbird, Britain's commonest land-bird. It must also be stressed that true polymorphism is a relatively rare thing in the avian world, the universal or practically universal type of variation being far subtler and less obviously marked and correlated only in a statistical way with the species' distribution.

The most striking and clear-cut of the ways in which the form of a population of birds or other animals marches with its environment are known as the ecological rules. The modern formulation of these rules owes much to the work of Rensch (1936 and many other papers). The ecological rules are set out very neatly by Mayr (1942, pp. 88-92). Those which apply to birds include Bergmann's rule which states that in a group of animals with a distribution extending through several latitudes or rather isotherms, the more polar populations will tend to be larger than the more tropical ones. Allen's rule states that protruding body parts, such as tails, ears, bills, extremities and so forth, are relatively shorter in the cooler parts of the range of the species, than in the warmer parts. A special case of Allen's rule may be the fact, first pointed out by Murphy (1938) that as a rule island birds have longer bills than related mainland races. Gloger's rule states that animals inhabiting warm and moist regions tend to have more dark pigment than their races in dryer and cooler regions, and that races in dry desert regions tend to have more yellow pigment. Special rules applying generally have been promulgated often by Rensch. Thus (1931) the alimentary canal of mixed feeders in the tropics is smaller than in temperate zone races of the same genus. There is a general correlation (Rensch, 1936; Rand, 1936) between wing-length and shape and altitude, the longer-winged, narrower-winged races inhabiting colder or higher regions. I suggest this should be called Rensch's rule, after its discoverer. Rensch (1938) also states (with an interesting table) that birds which live in the cooler part of the range of the species lay more eggs per clutch than the races in the warmer parts of the range. This idea was first put forward clearly by Hesse (1922) and can be known as Hesse's rule. Lack's (1947b) more recent exploration of this rule has shown that the trend of increasing clutch-size within a species runs latitudinally from the tropics, and that some of the European species which show this latitudinal trend also show a west-east trend of increasing clutch-size. He correlates the clutch-size with day-length, considering it to be ultimately determined by the maximum number of young for which the parents can find enough food; the amount of food the parents can find depends very largely on the length of day, though of course also on other factors.

The final rule for birds as first stated by Mayr (1942) is that races in the cooler climates are often more strongly migratory than the more southerly races. This could be called Mayr's rule.

Bergmann's rule works very well with a number of species which have a wide latitudinal breeding range extending into the arctic. Thus, the puffins, guillemots, eider-ducks and ptarmigan are all considerably bigger in the northern part of their range than in the southern. The ptarmigans also tend to be bigger on high mountains than on low mountains at comparable latitudes. Rensch has found, indeed, that among the races of resident birds of the Palearctic region, only eight per cent show exceptions to Bergmann's rule, though in North America twenty-six per cent are exceptions. Bergmann's rule, of course, depends on the principle that heat is lost relatively more slowly from a large body than from a smaller one of the same kind.

Gloger's rule has been found to be widely applicable. Rensch (1936) investigated sixteen species of the tit and nut-hatch family, and found only one exception to Gloger's rule. It generally holds good of birds of deserts and their more humid surroundings. The desert birds of the Sahara are much yellower and lighter than their coastal forms; the weavers of the Australian coast are much darker than their neighbouring desert relations. The many investigators of the complex races of the song-sparrow in the United States have found almost everywhere darkness marching with humidity, lightness with dryness. In Africa correlations between plumage and humidity are specially rewarding. Among the latest and most important papers has been that of Eberhard (1948) on the influence of climatic humidity on the plumage colour of the grey shrikes, *Lanius excubitor*, of North Africa.

In 1939 J. S. Huxley first used the word *cline* to recognize the situation where a species through its range followed an ecological rule. Of course the existence of clines had been known to systematists for a long time before Huxley thought of this useful word. Clines are gradients in space. They can be morphological gradients, gradients in size or in proportions or in the amount or intensity of one or more pigments. Some clines are very well marked and grade smoothly between widely separated extremes. When Huxley proposed them it was thought that the statement of their existence in some symbolized form by systematists might save the naming of a number of unnecessary races. But it must be said that so far there is no sign that naming has abated as a consequence. There may be clines also in the relative number of phases in the various populations of dimorphic or polymorphic species; these are known as polymorph-ratio clines.

The literature now contains many cases of ideal clines in which there is a regular and continuous gradation over the full distributional area of the species. In his own examples, chosen for the development in 1942 of his ideas of geographic speciation, Huxley did not restrict himself to such ideal clines, but quoted cases in which the gradation was either in the form of a stepped ramp or in an irregular form.

Writing at the same time and independently, Mayr (1942) put his finger more firmly on the difference between a species with a true cline and a species whose character-distribution is irregular. It is from the irregular, or at least not smoothly continuous, character-gradients that we can trace the origin of species. Mayr never wrote a truer word than when he stated "The more [smooth] clines found in a region, the less active is species formation".

Before we turn, then, to species formation, let us examine a few cases of truly continuous clines, or at least clines which are apparently so. Although the breeding distribution of the puffin, *Fratercula arctica*, has been grouped for convenience into a chain of subspecies, the size gradient of the species from France to the high arctic is a pretty smooth cline. There are large-scale and apparently smooth colour clines among all the chief species of silver pheasants (Beebe, 1921). Over at least a substantial part of the range of the common nuthatch in Europe Løppenthin (1932) finds a continuous cline in the colour of the underparts from chestnut to pure white extending for several hundred miles from west to east. But at each end the species passes out of cline and into regions of stabilized form. Snyder (1935) finds a continuous colour cline in the sharp-tailed grouse from north to south over the great plains of North America, though this is superposed on a group of sub-species marked by other (discontinuous) characters.

The rise in the number of bridled guillemots from south to north in the eastern Atlantic population gives a general impression of a smooth cline, though close examination discloses certain anomalies, and even what may possibly be reverse trends (Southern, 1951).

Mayr (1940) shows that among the birds of paradise, *Paradisea apoda*, of eastern New Guinea, there is a gradient in the amounts of two colours, the amount of yellow on the back increasing while the intensity of red in the plumes decreases. Mayr thinks it is probable that the two processes are independent of each other, since the entire change of the back colour occurs in the red-plumed birds and the entire change of the plumes in the yellow-backed birds.

Some truly smooth clines have been recently identified in European birds by Voous (1949). For instance, in several characters of the western European bullfinches and great and lesser spotted woodpeckers a distinct clinal variation (less prominent in the bullfinch than in the woodpeckers) takes place from north-west Italy and southern France to Denmark. Voous found the gradation so gradual that it was "hardly worthy of nomenclatural notice", which was his way of saying that he could not break the cline anywhere to form discrete subspecies. Among the European jackdaws Voous (1950) also found some clines in wing-length that sloped smoothly through the populations of what, on colouration, were several valid and discrete subspecies. Among the European jays (1953) he found separate smooth clines in colouration of upper-parts, colouration of under-parts, and size; these made it very difficult to recognize discrete subspecies, and he had to break the "seventy-five per cent rule" (see p. 99) to do so.

Among the drongos (Mayr and Vaurie, 1948) which inhabit India and the East Indies, the size, pigment, proportions and special structures such as crests and tail-racquets all vary geographically. Leaving aside the discrete island forms, typical clines occur in all the widespread continental species. Some of them are of special interest. For instance, a marked cline in crest-length in *Dicrurus paradiseus* is probably due in large measure to differential growth superimposed on an ordinary size cline conforming to Bergmann's rule.

It must be said that we have been adhering to the notion of an idealized smooth and continuous cline somewhat for the purposes of argument. There are, as we have seen, numerous clines which can defy the most ardent splitter to separate discrete subspecies out from them. But if we were to examine a cline, not population by population so much as bird by bird, we should find that we preserve uniformity only in the aggregate and statistical sense. The individuals differ from each other. Now wherever individuals differ, it is possible that populations may come to differ. It needs but the slenderest barrier between two groups in what was previously a continuous cline for those groups to begin a trend towards genetic differentiation. Thus the first stage of speciation is the dropping of modest barriers across a cline, and a consequent tendency to isolation, with consequent genetic reorganization. The characteristics of what was once a smooth cline may begin to arrange themselves as a stepped ramp, or the populations (each within its relatively isolated area) may develop smooth clines across their own areas while the transition from one population to another across the barriers would be more sudden. The point is that when discontinuity develops, a chain of populations is formed. By modern standards we can recognize these populations as valid subspecies and give them names if with regard to any token character seventy-five per cent of one population differs significantly from ninety-eight per cent of the population with which it is compared (Amadon, 1949; Rand and Traylor, 1950; Mayr, 1951). Mayr emphasizes that for practical and obvious reasons the character must be a taxonomic (or systematic) one, that is, recognizable in museum specimens. It must be the earnest wish of many confronted with the plethora of British and European 'subspecies' that a seventy-five per cent rule had come into operation by general consent twenty years ago.

The transition from subspecies to species comes when differentiation has progressed so far that when two subspecies find themselves in the same area (or are introduced into it) they are able to exist side by side without interbreeding, and without competing for the same food. Each must therefore have some mechanism for reproductive isolation, and a preference for some different food spectrum. For if they interbreed, all the work of gene recombination and genetic isolation may be rapidly undone, and if they compete for exactly the same bands in the food spectrum, one will drive the other out.

Two subspecies originally of the same species cannot become separate species unless in isolation they have become pre-adapted with isolating mechanisms and compatible ecological habits, particularly feeding habits. But a complete pre-adaptation in these respects is not necessary. It is for pre-adaptation to insert only the thin end of the wedge, which becomes broadened by further isolating mechanisms and ecological compatibility-factors (as Mayr calls them) by natural selection after the overlap.

If this method of geographical speciation be admitted, it must be the case that at any one time various kinds of animals are differentiated into non-overlapping subspecies which, were they suddenly brought into apposition, would be found to have already become separate species. These non-overlapping (recently-emerged) species are known, as we have already mentioned, as allopatric species. It is obvious in a large number of cases that it will be particularly hard to say (at least, without experiment) whether separate populations, on the invisible threshold, are good subspecies or allopatric species. It will thus never be possible to arrive at a final census of species. In such borderline cases a saving and useful clause has been introduced by the word superspecies, which is approximately the same as the *Artenkreis* of Rensch (1929). Really, the name superspecies is the name for a group of truly allopatric species directly descended from a single species.

To indicate the tenuous threshold between species and superspecies it is necessary to give some examples. These can be found in many orders and families of birds. In the view of Mayr and Amadon, for instance, there are only six species of pelicans in the world. In the view of Peters there are eight; for what Peters views as species members of a superspecies, Mayr views as subspecific members of a species. For similar reasons Mayr finds one snake-bird or *Anhinga* only in the world with four subspecies, while Peters regards them as four allopatric species. Other very good examples of borderline cases are found among the geese, of the New World the Canada geese, and of the Old World the bean-geese. The case of the bonxie or great skua is an interesting one. One hypothesis to account for the present distribution of this bird (though not one which can be proved) is that the species *Catharacta skua* represents a more ancestral form than the other three smaller skuas, *Stercorarius*, of today. Skuas are probably descended from gull-like birds, and it is probable that one 'discovered' the Antarctic regions before the *Larus* gulls which originated in the north. Here it filled the ecological niche of scavenger and predator which would have otherwise been filled by big *Larus* gulls, and either became or remained more gull-like and primitive in habits than the other skuas. Much later, and quite likely in recent times, one of the races of the southern bonxie *C. s. skua* wandered back into the North Atlantic and settled and bred in an area ecologically most resembling that of the sub-antarctic islands from which it had come—namely, Orkney, Shetland and south-east Iceland, the only places where this bird is found nesting in the northern hemisphere. Most people agree that the differentiation of the bonxies, including the North Atlantic subspecies, has not yet reached species level, though if an alleged breeding overlap in British Antarctica (p. 103) proves to be true, the group will have to be raised to superspecific status.

The only other bipolar bird, the fulmar, has been generally agreed to be a superspecies. Peters in 1931 placed the antarctic fulmar not only in a different genus but even in a different subfamily from its northern member. It was Voous (1949b) who first pointed out the extraordinary similarity between the antarctic and northern fulmars, the chief difference being only in the shape and length of the bill. It seems probable that the northern fulmar colonized the northern hemisphere from the south, and it subsequently became differentiated, perhaps by the Pleistocene glaciation, into two different subspecies.

It would be a matter of no difficulty for a competent systematist to identify superspecies formed by groups of species of birds of any of the temperate or tropical continents. Mayr's years of work on the birds of the East Indies produced many such examples in his papers. The superspecies among the ducks are discussed by Delacour and Mayr (1945). Quite a number can be constructed among the pigeons and doves. The collared turtle-dove *Streptopelia decaocto*, now spreading so remarkably in Europe, belongs to the same Artenkreis as an East Indian species, and an important group of half-collared African doves centring round *Streptopelia roseogrisea*, which is probably also the ancestor of the domestic Barbary dove.

Some cosmopolitan birds such as the barn-owl are highly differentiated into groups of species and allopatric species. Thus *Tyto alba* has at least thirty-four valid subspecies from every part of the temperate and tropical world except New Zealand, and a few islands, some of which are inhabited by allopatric species such as *T. rosenbergi* of Celebes.

We have followed the species' differentiation from the situation where the characteristics of its populations are distributed in an idealized series of clines, to the stage where the populations tend to become geographically isolated, and what were idealized clines become stepped ramps. At some indefinable stage in this process, speciation occurs, and one set of geographical subspecies becomes two or more sets of allopatric species, each of which may of course include several geographical subspecies. This is the stage of the superspecies. The genetic difference between species belonging to a superspecies will (to recapitulate) be caused by three kinds of isolating mechanisms; those which are neutral, those which are preadapted and those which are acquired by selection, and three corresponding kinds of ecological compatability factors (Mayr, 1951).

The member species of the superspecies undergo a period of reorganization in isolation. The test of whether the member species are truly allopatric species and not well-marked subspecies can only come in the process of history unless man takes a hand and consciously or unconsciously introduces one form into the range of another. Sooner or later history will take its course and the range of one form, changing as ranges always are, will impinge upon the range of another. To use the language of the bull-ring, this is the 'moment of truth'.

If at the moment of truth the genetic differences are not enough, there will be a fairly swift period of hybridization resulting in reintegration. It is difficult to find examples of this process, which must theoretically be fairly common in nature; but a lot of the difficulties systematists have had in sorting out the yellow wagtail, *Motacilla flava*, is probably due to the presence of several such zones in which the races of this interesting and variable species are at present reintegrating. (Domaniewski, 1925; Stresemann, 1926; Johansen, 1946; Smith, 1950.)

Another species which may be at present undergoing the joining-up test is the bonxie or great skua, *Catharacta skua*, already mentioned. In the sub-Antarctic South Shetland islands both the subspecies *C. s. maccormicki* and *C. s. lönnbergi* may be present. The evidence is not sufficiently clear on whether they breed discretely or whether they hybridize (Murphy, 1936). It is possible that we are here observing the incipient speciation of a form long differentiated subspecifically. Meinertzhagen (1952) suggests the name 'rejungent' species for forms which, hitherto regarded as subspecies of the same species, have met again but differ to such an extent that they are reproductively isolated. He thus applies the word to the hitherto allopatric species which have newly become sympatric. He goes on, however, to suggest that because of their recent common ancestry, rejungent species should be regarded as still conspecific, and with this many must find it impossible to agree. He gives many very valuable examples of species-pairs which have stood the test of overlap. There is the classic case of our great tit, *Parus major*, and its Chinese form *Parus minor*. These have been shown by Rensch (1933) to overlap in the Amur River region into which the *Parus major*, from his evidence, made a post-Ice Age eastward extension. Although connected by a more southerly chain of subspecies involving the intermediate group *bokharensis*, the two forms where they overlap do not interbreed but live side by side as good species.

The classic example of speciation at the end overlap of a chain appears in every textbook of systematics. It is that of the medium-large gulls of the genus *Larus*, exemplified in Britain (which is in the area of overlap) by the species *Larus argentatus*, the herring-gull, and *Larus fuscus*, the lesser black-back. The chain is branched as well as overlapping, and has been described at various stages of the knowledge of the systematics of this group, by Stegmann (1934), Mayr (1940), Stresemann and Timoféeff-Ressovsky (1947) and Fisher and Lockley (1954).

In his exciting systematic review of the genus *Phylloscopus*, C. B. Ticehurst (1938) shows that the greenish warbler, *P. trochiloides*, has two subspecies which meet in the Altai mountains, where they do not interbreed. They are connected by a cline-form chain of subspecies. Among the chiffchaffs, *P. collybita*, there is also a probability that the two subspecies that meet in the western Caucasus do not interbreed, and are therefore sympatric species. Another well-known example of chain-overlap is the case of the hooded crow, *Corvus cornix*, and the carrion-crow, *C. corone*. This is one of the cases where overlap has taken place at what must be the 'point of balance' between subspecies and species, with the result that there is a fairly stable hybrid zone. Mayr suggests that there is no evidence that that of the crows has 'broadened materially' within the last 5,000 years, and he shows that such cases are not very uncommon. Besides the crow case, which was described first in detail by Meise in 1928, there are similar cases among nuthatches (see Voous and van Marle, 1953), long-tailed tits, shrikes, goldfinches, black-headed buntings, in the New World grackles, flickers and juncos; and, in the Indian region, particularly silver pheasants. Most of these hybrid zones owe their existence to the breakdown of some former geographical barrier. It seems probable that if the crow case has been in balance for as long as Mayr suggests, it may be the exception that proves the rule that in most cases the populations in these zones must shortly integrate or differentiate. That most interesting and remarkable book of Lack (1947) on Darwin's finches of the Galapagos Islands gives descriptions of similar cases.

So far we have considered cases in which the populations have been more or less continuously distributed over large areas of land. But much speciation appears to have taken place as a result not so much of creeping overlap as of jumping overlap. Sometimes the jumps have been to islands, and sometimes across oceans to new countries and even to new continents. The possibility of the colonization of countries by species pre-adapted to live in them, across immense distances of barrier, have been neglected in most recent literature, and it is only now that the possibilities of 'drift dispersal' are coming to be understood, largely through the work of Finn Salomonsen in Denmark and Kenneth Williamson on the Fair Isle, who suggest explanations for such things as the frequent appearance in Western Europe of passerines not known to breed west of Siberia. The transatlantic crossings of many American passerine species to Britain (which have been steadily and stolidly recorded as 'probably escaped from captivity' in the British check-lists) and the spectacular colonization of Greenland in 1937 by the fieldfare, *Turdus pilaris* (Salomonsen, 1951b)—in short, the possibilities of invasion, are only beginning to be realized.

And yet, staring all observers in the face, are the undebatable cases of small oceanic islands in many parts of the world isolated by hundreds, sometimes by thousands of miles of sea; yet supporting populations of one, several, and in a few cases many indigenous, specialized and peculiar birds descended from chance drifted species, and in many cases so different that their real ancestors can only be guessed at.

But to consider first the invasion of continents, the famous *Larus* chain gives us our first example of transoceanic colonization, for the black-backed end of the chain, resting on Britain, was overlapped across the Atlantic Ocean by the herring-gull —not necessarily *via* Iceland for there is evidence that Iceland's herring-gulls have colonized that country only fairly recently.

The north-western part of North America is inhabited by a subspecies of our common gull, *Larus canus*, known as the short-billed gull. The ring-billed gull, *Larus delawarensis*, may be the eastern North American representative of what was originally the same species, colonized from the Old World across the Atlantic, and subsequently differentiated in isolation. The ranges of the two now overlap. Another case may be that of the pochards, the ducks of the genus *Aythya*. In the Palearctic there is but one common pochard, *Aythya ferina*. In North America the two species, *A. valisineria* (the canvas-back) and *A. americana* (the redhead) replace the Old World's single species. They are well differentiated from each other, but it is possible that each is a transocean colonist of a different side of the New World, originating from the Old.

All islands situated in the remotest areas of the sea appear to receive land-bird visitors at intervals[1]. It is often, indeed usually, a matter of chance as to which species is the first to get there, and it is clear that many such island species have had time to differentiate, so that when a subsequent further colonist of the same mother species arrives, the descendant of the original will be sufficiently distinct to remain separate. Cases of double invasion of this kind are now well known, and there are even some cases of treble invasion—for instance, the existence of three different species of *Zosterops* or white-eye on Norfolk Island, east of Australia, each of which is probably a direct descendant of the same mainland species.

Mayr (1942, p. 173) listed a series of very beautiful cases of species-pairs on islands derived from double invasion by the same mainland species. From the Canaries, for instance, the two chaffinches, *Fringilla teydea* and the local Teneriffe race of *F. coelebs*, and the two pigeons *Columba laurivora* and *C. bollii*, from Samoa the caterpillar-eaters *Lalage maculosa* and

[1]G. R. Williams (1953) has shown how surprisingly frequent island visits can be. Of the many British passerines introduced to New Zealand and Australia, mainly between 1860 and 1870, thirteen are now established in New Zealand and eight in Australia.

Nine islands or island-groups (Sunday, Chatham, Antipodes, Campbell, Snares, Auckland, Macquarie, Lord Howe and Norfolk) lie at distances of 200 (except for Snares, 60 miles from New Zealand) to 550, average 360 miles from Australia or New Zealand, whichever is the nearer. Within thirty or forty years of their liberation many of the passerines began to appear on these islands, and from one to ten species now breed on each. Most must have been drifted by strong winds from New Zealand, and the starling now breeds on seven and has been seen on all nine.

L. sharpei, from Ceylon the woodpeckers *Brachypternus erythronotus* and *B. benghalensis*, from Luzon the pittas *Pitta kochi* and *P. erythrogaster*, from Celebes the drongos *Dicrurus montanus* and *D. hottentottus*, and in Tasmania two species of thornbill, *Acanthiza pusilla* and *A. ewingi*. There are many other examples. Thus the two finches of Nightingale Island, belonging to the extraordinarily isolated group of Tristan da Cunha, *Nesospiza acunhae* and *N. wilkinsi*, are probably derived from separate invasions of the same mother species. But the best examples of double or several invasion species come from the two most biologically isolated archipelagos in the world, the Galapagos and the Hawaiian Islands. In particular, the study of the Darwin's finches of the Galapagos by Lack (1947)—also that of the honey-creepers of Hawaii by Amadon (1950)—have disclosed situations of the utmost complexity and fascination. In these archipelagos the invasions have been not only from mainland to archipelago, but from island to island within each archipelago. Each Galapagos island supports between three and ten species of Darwin's finch, and there are fourteen species in all, one of which lives on the isolated island of Cocos. Two of the forms collected by Darwin in 1835 have since become extinct. It was this first encounter with the family Geospizidae that probably turned Darwin's thoughts to the theory of organic evolution. The Darwin's finches of the Galapagos have reached their high degree of specialization and complexity over a period sufficiently long for their original South American mainland ancestor to have disappeared; at least, there is no existing mainland species which can be reasonably and credibly suggested as the modern representative of their ancestor. All the Darwin's finches are isolated from each other ecologically, and the degree of adaptive radiation is considerable. Lack divides the group into the genera *Geospiza* (six species), *Camarhynchus* (six species), *Certhidea* the warbler-finch and the Cocos finch, *Pinaroloxias*. Four of the *Geospiza* species are ground-living seed-eaters, but two are specialized for feeding on cactus. Of the tree-finches, *Camarhynchus*, one is vegetarian, three insectivorous, one fills a woodpecker niche, using a stick to probe for insects instead of a long bill, and the sixth is associated with mangroves. The warbler-finch *Certhidea* is a ground-finch which has evolved a real insectivorous warbler-like form.

In these interesting island studies of Lack, the ecological differences between the members of the groups of closely allied species are the significant features. Lack's investigations have served to accentuate and bring out the fact that no two members of one of these groups of species eat the same food. Amadon's studies of honey-creepers have shown the same. Each bird can be shown to have a different food spectrum, to have different food preferences, and a different staple diet.

Species-pairs and species-groups are not, of course, confined to islands where there have been two or more invasions. Mountain ranges, especially high and isolated ones, produce 'island conditions' and there are many cases of double-invasion species to be found in mountains. And, of course, species-pairs can equally be rejungent species at points of overlap in subspecies chains where the overlapping members are found to have speciated. It is, then, to the study of species-pairs and groups of closely allied species that the new ecological systematist can most usefully turn.

So far, every examination of a true species-pair has disclosed a fascinating ecological divergence. But there are many well-known species-pairs whose general habits are known to be different, but whose food ecology has never been properly explored. Such are, rather oddly, the herring-gull and the lesser black-back, the two very closely allied species of tree-creeper in Western Europe, *Certhia familiaris* and *C. brachydactyla*, the marsh- and willow-tits, *Parus palustris* and *P. atricapillus*, the arctic, common and Forster's terns, *Sterna macrura*, *S. hirundo* and *S. forsteri*, and the nightingale and sprosser, *Luscinia megarhynchos* and *L. luscinia* (Stresemann, 1948). These species overlap in a well-marked area between the Vistula and the Oder, and the sprosser has spread very considerably westwards in the Upper Danube and to one area on the Elbe in the last 200 years. In captivity they will inter-breed but in the wild their ecological divergence seems quite great enough to keep them from doing so, and they have several behaviour organs, such as song, also adapted for this purpose. Unfortunately, nothing seems to be known about the fundamental differences in their food. Even such interesting species-pairs as the common and Brünnich's guillemots, *Uria aalge* and *U. lomvia*, have not had their food differences intensively studied in the areas where they overlap. But thanks to their putative economic importance, the species-pair cormorant and shag have had their food analysed, and found to be entirely different (Steven, 1933; Lack, 1945; Lumsden and Haddow, 1946), and the intensive analysis of North American wildfowl food by the U.S. Fish and Wildlife Service, also performed for economic reasons, has disclosed that every member of a species-pair in this group of birds has a rather well-marked food difference (see Bent, 1923, 1925; Kortright, 1943; Cottam, 1939). Thus the redhead, *Aythya americana*, eats a considerable amount of musk-grass and rather little wild celery, while its sympatric pair-species the canvasback, *A. valisineria*, eats considerably more wild celery than musk-grass, though both eat more pondweed than anything else. The lesser scaup, *Aythya affinis*, eats more vegetable food than does the greater scaup, *A. marila*. The lesser scaup eats more grasses than the greater, the greater more musk-grass and water-milfoil than the lesser. The greater eats a good deal more molluscs than the lesser, the lesser more insects than the greater.

The whole problem of interspecific competition is well summarized by Udvardy (1951), who, carefully examining the birds of Europe, finds that: "So far no populations of differing species which compete for food are known upon the European continent"—except the populations of different forms which are actually in the course of speciation. Udvardy concludes that:

> "The European continent is not yet saturated with bird species and can offer possibilities of existence to numerous new forms through immigration or speciation without risking competition worth mention with the species already living there."

CHAPTER VIII

ABSOLUTE NUMBERS OF BIRDS

IN *Birds as Animals*, published in 1939 (p. 92), I quoted only one instance of a bird species whose world population was known, the gannet, and one subspecies, the St. Kilda wren. If I had been better acquainted with the literature then I would no doubt have been able to quote a few more cases. But there is no doubt that our knowledge of bird numbers has advanced a great deal in the last fifteen years, and there are now numerous forms of which the world population is known with some accuracy; not all of them are rare birds whose population is largely determinable because it is small. All the species with which I can illuminate this chapter are of birds whose population has been counted or estimated by some direct method, and not by comparison with the numbers of some other species, or by the indirect methods described in the next chapter, which is concerned with the numbers of the different elements in bird communities.

In 1939 I suggested that it would be difficult to establish the world populations of any birds at all accurately except under one or more of the following conditions:—

(a) that they are social sea-birds with a limited number of colonies, and a definite breeding season, so that expeditions can count the number of occupied nests at each colony.

(b) that, if land birds, they are social birds with a restricted geographical distribution and a limited number of colonies.

(c) that they are large and spectacular birds, easy to recognize, which are nearly extinct or which have a relatively small breeding area in which the territory or hunting ground of each individual pair can be discovered.

(d) that they are species of birds restricted to small islands.

Although I have been proved right in so far as world censuses have now been made of birds in all these categories, it is now clear that I was unduly pessimistic. For a world census of the rook I now know to be a practical, though expensive, possibility, and the rook has a very wide distribution and an extraordinarily vast number of colonies which make it antithetical to category (b). By the same token, a world census of the stork, at least in the western half of its distribution, has practically been done, and the stork is a large, spectacular and easily recognized bird which could be included in category (c) were it not far from extinct, and widely spread.

There are two further and important methods of estimating wild populations of whose existence I was quite unaware in 1939, though one of them had been proposed by F. C. Lincoln as long ago as 1930. The principle of the Lincoln index is a simple one. If a sample of a population is caught, marked, released and subsequently completely mixes with the population from which it was taken, then the proportion of marked birds in a subsequent sample can lead to a direct calculation of the size of the original population sampled. Thus, let us suppose that a bird population *P* was sampled, and that 100 of its members were caught, marked and released, returning to the population from which they were taken and mixing themselves with it. Let us next suppose that subsequently another sample of 100 was taken at random from it, and that of this 100, ten were found to be marked, that is, one-tenth of the sample. It would be thus possible to conclude that as one tenth of the sample were proved to have been previously marked, the previous operation had succeeded in marking the same proportion of the entire population *P*. The first sample of *P* was 100 birds; thus the value of *P* was ten times 100, e.g. 1,000.

All sorts of corrections to this idealized and over-simple calculation have to be made: corrections for the fact that in between samplings the birds may reproduce themselves and some will die; corrections for the fact that no sample taken from a population ever mixes entirely with that population when it has returned, and that (in a sense) no sample of a bird population is a random one, in that families or groups from small regions tend to stick together. Nevertheless, developments of the Lincoln index system have been successfully used to estimate the world population of certain species of wildfowl, notably of the pink-footed goose.

The second new method of determining world populations has also successfully been developed in the study of wildfowl, notably by Peter Scott in Britain, and in the U.S.A. by J. D. Soper (1942) and by Hanson and Smith (1950). These workers have begun to determine the total populations of several species of wild geese by the collecting of counts taken close together or simultaneously by a network of co-operators, of every winter flock. So well are these now known in Britain and the eastern United States that no substantial weather movement of geese from one part to another can escape notice, and neither are any geese but a few individual stragglers likely to escape notice. Soper's winter flock estimate concerns what is probably the world's most numerous goose, the blue goose, *Anser caerulescens*. The evolutionary status of this most interesting form is in some doubt. There is a lot of evidence that it is a colour-phase of the snow-goose, and in at least one of its breeding grounds it freely hybridizes with the smaller and much more numerous of the two snow-goose subspecies, the lesser snow-goose, *Anser hyperboreus* (*caerulescens*) *hyperboreus*. Incidentally the greater snow-goose, *A. h.* (*c.*) *atlanticus*, the large northern form, has, according to Scott (1950) a world stock of about 17,000, apparently increasing; for it was estimated at about 12,000 in 1938. Most of the lesser snows migrate to the western United States. But the blue geese all go to the Mississipi Delta, and places west into Texas along the coast of the Gulf of Mexico, where they are accompanied by lesser snows to about one-tenth of their number. In a most careful survey of the literature, and of the important observations of E. A. MacIlhenny (1932), Soper "feels impelled to the conclusion that the grand total of blue geese in existence may not be less than five million." In spite of frequent references to such a thing in the non-ornithological literature, birds seldom darken the sky; but blue geese do, and Soper prints some photographs which show that they do.

No other wild goose with a comparable population to that of the blue goose has been estimated. But, judging by the size of winter flocks, Scott (1950) suggests that the population of the yellow-billed Greenland subspecies of the white-fronted goose, *Anser albifrons flavirostris*, was "possibly between

20,000 and 30,000". These can be fairly easily assessed, as, with the exception of a few erratic individuals, they all winter in Ireland. Another subspecies of the white-front, the tule goose, *A. a. gambeli*, has a much more restricted breeding and wintering distribution. It is the largest of its species, breeds only in the Perry River area of the Canadian Arctic, and winters only in a restricted area of the Upper Sacramento Valley in California, where its numbers may be "no more than 1,000". It is in the Perry River area of the Canadian North West that breeds the smallest of the snow-goose subgenus, *Anser rossii*, the little Ross's goose. After the breeding grounds of this goose had been discovered in 1940, but before his own visit in 1949, Scott, from his knowledge of the winter flocks, which are restricted like the tule goose to central California, estimated that the world population was "probably between 2,000 and 3,000". After his expedition of 1949 he showed (1951) that the stock in the Perry River area amounted to not more than 2,000 individuals, including the 1949 season's goslings, and so far no other breeding area is known. Ross's goose and the pinkfoot, *Anser brachyrhynchus*, are so far the only two species of geese whose world population has been tackled in both wintering and breeding grounds. In the case of the pinkfoot the attack on the problem has really been threefold, for its numbers have been investigated (Scott and Fisher, 1953) by the estimation of winter flocks, by that of the summer breeding population, and by sampling on both wintering and breeding grounds, with marking and with the subsequent application of the Lincoln index. All these methods show that the population since 1949 has been rather consistently between 30,000 and 40,000 geese and goslings each October. A big toll, amounting to about twenty or thirty per cent of the winter population, is taken every year by wildfowlers, but the species can evidently stand this cropping and recover its numbers adequately on the breeding grounds. These, incidentally, were not fully known until the Severn Wildfowl Trust expedition to central Iceland in 1951, where the chief breeding place of the species was discovered, and thoroughly investigated, no less than 1,151 geese and goslings being caught and marked when the adults were in their four-weeks' flightless moult. In the succeeding winters about twice the number were caught in England and Scotland in rocket-propelled nets, so that by the winter of 1952 more than one in every ten individuals of the wildest of our grey geese were flying about with rings placed on their legs by Scott. More than half the wild population of the pinkfoot is now known to nest in vegetation oases in the true tundra of the head waters of Iceland's longest rivers, the rest nesting in a couple of dozen valleys of Spitsbergen, and over a long stretch of the high arctic east coast of Greenland. The pinkfoot population winters almost exclusively in England and Scotland, the only substantial flock in any other country being a group of about 4,000 birds on the North Sea coast of Germany.

In 1953 Scott visited the central Icelandic breeding-ground of the pinkfoot again, and his expedition caught no less than 9,005 different geese and goslings: by the winter of that year over a sixth of the living world population had been marked!

Many authorities hold the pinkfoot to be a subspecies of *Anser arvensis*, the bean-goose. But Scott regards it as a good allopatric species, and its distinction is a very clear one not only in field marks but also in many aspects of behaviour. It is certainly a better species than the rare trumpeter-swan of America, which Scott regards as a subspecies of the whooper swan, *Cygnus cygnus*, in which the yellow of the bill has been replaced by black. Incidentally, he regards the present population of this rare bird, the trumpeter-swan, *Cygnus cygnus buccinator*, as being between 1,000 and 1,500, though ten years ago it appears to have been considerably less.

The barnacle-goose, *Branta leucopsis*, a black goose which has a very restricted high arctic breeding distribution in north-east Greenland and Spitsbergen, winters mainly in Ireland, western Scotland, Holland, Denmark and the western end of the Baltic. This is a decreasing species which according to Scott (1950) has a world population "probably less than 30,000, decreasing". But the rarest known surviving goose is the ne-ne, *Branta sandvicensis*. Scott (1953) shows that in June 1952 there were at least thirty-three, but probably not more than fifty individuals of the species still alive in the Hawaiian archipelago, of which twenty were in captivity or semi-captivity, that twelve were in the collection of the Severn

Wildfowl Trust in Gloucestershire, of which nine were goslings reared in 1952 from the gander and two geese with which Scott started his 'restoration-experiment'. In 1951 the total number of the species was unlikely to have exceeded fifty, and it seems that, like those of Père David's deer (a mammal of Chinese origin now confined to Bedfordshire, England) the fortunes of the ne-ne may be entirely restored by breeding in captivity.

Since ducks do not form winter grazing flocks like geese, their numbers have not been estimated in the same way, and we can only guess at the world population of a few rare species or subspecies with restricted range. Thus, the Hawaiian duck, *Anas wyvilliana*, which is often regarded as a well-marked subspecies of the mallard *A. platyrhynchos*, may only number between 300 and 500 individuals (Scott, 1951b). A related subspecies, or species, *A. p. oustaleti*, may not number much more than a couple of dozen individuals (R. H. Baker, 1951), and yet another in this group of allopatric species or well-marked subspecies of the Pacific, *A. p. laysanensis*, the Laysan teal, is probably now extinct. So may be another island duck, the merganser, *Mergus australis*, of the Auckland Islands, 400 miles south of New Zealand.

When we consider the world population of certain sea-birds, we find, not unexpectedly, that the only practical method of estimation has been, so far, the discovery of all colonies and the determination of the number of occupied nests at each, if possible in the same season. Only at places where sea-birds are of economic importance, such as on the guano islands of Peru and Chile, has there been any attempt to estimate the numbers of species of great abundance—for this is an excessively difficult task. The great guano industry of the Humboldt Current has been described by Murphy (1936), Vogt (1940–42) and Hutchinson (1950). The three principal species, which produce an annual guano deposit of 100,000 tons or more, are *Phalacrocorax bougainvillii*, the guanay cormorant, *Sula variegata*, a booby known as the piquero, and *Pelecanus occidentalis thagus*, the big Peruvian subspecies of the brown pelican, in that order. In some years the total sea-bird population of these islands in the Humboldt Current certainly reaches seven million and probably ten, and the commonest species, the guanay, may number nearly 4½ million in a bad year (so Vogt, 1940, calculated from an aerial photographic survey), and may reach about 6 million birds in a good year. It is not possible to determine from the literature what the number of piqueros may amount to, but it is obviously over a million, though probably under 4 million. The number of pelicans probably does not exceed a few hundred thousand.

While the guano birds of western South America may be the most numerous sea-birds whose total numbers have been guessed at, they are probably not the most numerous sea-bird species in the world. Elsewhere I have more than once made the suggestion that Wilson's petrel, *Oceanites oceanicus*, of the Antarctic and sub-Antarctic, may be the world's most numerous species. In the higher latitudes of both north and south the abundance of plankton supports vast populations. The sooty tern, *Sterna fuscata*, is probably the most abundant tropical sea-bird. Many of its Atlantic, Indian Ocean and Pacific colonies number hundreds of thousands of birds. Although its normal clutch is one egg, 120,000 eggs have been gathered in a week on Ascension Island. On several islands north of Madagascar there are vast colonies of this species. On one, about a quarter of a million young hatched after a very heavy egg-collecting campaign and at one time in the present century it was estimated that five million pairs nested on the island. In Audubon's time there may have been more than a quarter of a million nests on the Dry Tortugas, the only sooty tern colony in the United States, and though the birds were down to under 4,000 individuals in 1903, they are back to over 100,000 (i.e. over 50,000 nests) fifty years later. The world population of the sooty tern must be many tens of millions, though perhaps not a hundred million. It is far greater than that of the bird of which Darwin (1859) wrote "the fulmar petrel lays but one egg, yet it is believed to be the most numerous bird in the world". I have shown (1952) that certainly over a million nests are occupied by the Atlantic subspecies, *Fulmarus glacialis glacialis*, each year, but I do not believe that the true figure will prove to be many millions and would be surprised if it were as much as two millions.

But more closely than that of the fulmar can we arrive at the world population of the great shearwater, *Puffinus gravis*, whose breeding places are none other than the islands of the Tristan da Cunha group, principally Nightingale and Inaccessible and their attendant islets. In the season of 1949–50 M. K. Rowan (1952) estimated between two and two-and-a-half million nests. She based this estimate primarily on the number of nests to the square yard on a sample of the shearwaters' ground on Nightingale.

The sea-bird whose population is perhaps most accurately known is the North Atlantic gannet, *Sula bassana*, whose history has been recounted by J. H. Gurney (1913), Wynne-Edwards, Lockley and Salmon (1936), and Fisher and Vevers (1943–44, 1951). Gurney was the first writer to realize that it was practicable to make an assessment of the number of nests occupied by this species. He suggested that in about 1912 about 55,000 nests were occupied at the colonies then known to him. This was a slight under-estimate of the gannet population at the time (which was probably near 65,000 nests) but it was a very careful effort to make the best of the information then available. Using counts and estimates at the known colonies made principally between 1928 and 1935, Wynne-Edwards, Lockley and Salmon subsequently arrived at a figure of about 78,000 occupied nests.

In 1939 H. G. Vevers and I tried to organize a census of all the nests in the one season. In that year the gannets were breeding at twenty-two known colonies, twelve in Britain, one in the Faeroes, three in Iceland and six in the Gulf of St. Lawrence. One small colony in France in 1939 was unknown to us; with this exception we managed to organize counts of the nests at all the colonies on the east side of the Atlantic and of two in the Gulf of St. Lawrence. The four other St. Lawrence colonies were counted between 1934 and 1941. 83,022 nests were counted in all, of which 69,782 were on the east and 13,240 on the west side of the Atlantic. In 1949 we once more tackled the gannets with the help of friends, though this time on the east side of the Atlantic only. We found twenty-three colonies, where previously there had been seventeen, new colonies having been established in the Channel Islands, Brittany,

Norway and Iceland. From about 69,782 nests the total rose to 82,394, an increase of eighteen per cent in ten years, and a clear continuation of the steady upward trend in the gannet's population since it appeared to be at its lowest on the east side of the Atlantic, at probably under 50,000 nests, in 1889.

The gannet has been much persecuted in the last hundred years, and its present increase is mainly due to protection. As this protection is being continued, it will be interesting to see what the limiting factor will ultimately be.

There are closely allied species of gannets in South Africa, *Sula capensis*, and in New Zealand and Australia, *S. serrator*. The populations of Africa and Australia, though high, have never been, as far as I know, the subject of an accurate census, but the whole group of the thirty-two New Zealand, known or suspected, gannetries were tackled by the Ornithological Society of New Zealand, particularly in the season 1946–47, organized by Fleming and Wodzicki (1952). They concluded that the number of occupied nests was between 18,000 and 24,000 and probably in the neighbourhood of 21,033. As with the North Atlantic species, they are intending to make further censuses at intervals of about ten years, and they also find the species to be increasing.

A relation of the gannet whose numbers have also been estimated by counting occupied nests at all the colonies is the American white pelican, *Pelecanus erythrorhynchos*, of which B. H. Thompson (1932) suggested there were 30,000 breeding birds, all in the U.S.A. and Canada.

The rest of the birds whose world populations can be approached seem to be all species with restricted distributions. The extraordinary Kirtland's warbler, *Dendroica kirtlandii*, whose breeding range is restricted to the jack-pine barrens of the state of Michigan and which winters in the Bahamas, as has been so interestingly described by van Tyne (1951) probably numbers no more than 1,000 individuals; Mayfield (1953) and his colleagues in the very thorough search of the whole area in 1949 could find no more than 432 singing males. The great white heron, *Ardea occidentalis*, which breeds only in Florida, may have been reduced (Peterson, 1950) to about 150 birds by a great hurricane in 1935, though now it may number over 1,500. The whooping crane, *Grus americana*, has not fared so well. In 1942 it may have numbered 200, in 1948 only thirty-two plus two others in captivity. The wild ones were thirty-one in Texas and one in Louisiana. Another rare American bird is the California condor, *Gymnogyps californianus*, of which there are perhaps not more than 120 in the world, in California and Lower California, of which about half (see Koford, 1953) live on the United States side of the border. Some species thought to be extinct have recently been rediscovered and numbered. The cahow of Bermuda, *Pterodroma cahow*, a rare gadfly-petrel, was never thought to be quite extinct, but its status was quite unknown till 1951, when Murphy and Mowbray (1951) found thirteen or fourteen nests and estimated the population as "perhaps of the order of 100 adult birds". The remarkable flightless rail of New Zealand, *Notornis hochstetteri*, was rediscovered by G. B. Orbell in 1948, having been believed extinct for more than fifty years, though we now know (Cumming, 1953) that one was seen in 1910. In the season 1949–50 the field workers estimated the population to be between forty and fifty birds (Falla, 1951) and since then a wider reconnaissance has shown that this may be a considerable under-estimate.

There remain other birds on the verge of extinction, and island forms. There may be only two female ivory-billed woodpeckers now left in continental North America and it seems uncertain (Bond, 1950) whether any survive in Cuba. Some restricted island forms enjoy more respectable populations; for instance, the extraordinary flightless rail, *Atlantisia rogersi*, of Inaccessible Island in the Tristan group, had probably between 10,000 and 20,000 individuals in 1950–52 (Elliott, 1953.) In 1941 Abbott's booby, *Sula abbotti*, now restricted to Christmas Island in the Indian Ocean, had between 500 and 750 breeding pairs (Gibson-Hill, 1948). I have already mentioned the duck, *Anas (platyrhynchos) oustaleti*, of the Marianas, which may number not much more than twenty-four. The monarch, *Monarcha takatsukasae*, of Tinian Island in the Marianas, also described by Baker (1951) was estimated by Gleize (1945) to be between forty and fifty in that year. Fleming (1939) shows that the interesting 'robin' of the Chatham Islands group east of New

Zealand is now restricted to not much more than an acre of jungle on the top of one tiny islet in the group, where there were not more than twenty to twenty-five pairs of *Miro traversi* in the season 1937–38. There are of course many of these restricted island forms, which have not been visited or found for many years, so that we do not know whether they have indeed any longer any population at all. Some of the full species quoted by Baker (1951) in this category from Micronesia include the Kusaie black rail, *Aphanolimnas monasa*, which has not been seen since 1828, the monarch *Metabolus rugensis* of Truk which has not been seen since 1931, the mountain-starling *Aplonis pelzelni* of Ponapé, which was seen in 1948 but is extremely rare, and the mountain-starling of Kusaie, *A. corvinus*, which has not been found since 1828.

From quite a different part of the world Bond (1950) gives comparative examples of extraordinarily rare species. Thus, in the Zapata swamp of Cuba live a rail, *Cyanolimnas cerverai*, a wren, *Ferminia cerverai*, and a finch, *Torreornis inexpectata*. These three species have a range of but a few miles and are found nowhere else in the world. Other rare and restricted species from the West Indies whose status is precarious include the dove *Leptotila wellsi* of Grenada, the mocking-bird *Mimus magnirostris* of St. Andrew's, the mocking-bird *Ramphocinclus brachyurus*, now extinct on Martinique and rare in Santa Lucia, the thrush *Mimocichla ravida*, restricted to the small island of Grand Cayman and last seen on it in 1938, and the warbler *Leucopeza semperi* of Santa Lucia. Similarly the pie chanteuse or Seychelles magpie-robin, *Copsychus seychellarum*, now survives only on Frigate Island in the Seychelles in the Indian Ocean, in which group also the Seychelles black paradise-flycatcher *Tchitrea corvina* survives only on La Digue, and the petit merle des îles, *Benrornis seychellensis*, is entirely restricted to one tiny island, planted with coconuts, of sixty acres only (Foster-Vesey-FitzGerald, 1953).

A search through the faunas would certainly disclose many other examples of species with extraordinarily low populations. Indeed, two or three species must still become extinct in each decade. This might seem rather serious from the protection point of view until it is realized that the majority are island forms and rather often allopatric replacement-species of other surviving species. Moreover, the losses of this kind are possibly counterbalanced by the number of new species emerging, crossing the still undefinable threshold that separates a well-marked subspecies from an allopatric species.

CHAPTER IX

COMPARATIVE NUMBERS OF BIRDS

IN the previous chapter we have been somewhat unhistorical; for the possibility of counting the absolute numbers of a world species was scarcely realized when bird census work was first taken seriously, rather more than fifty years ago. It was not until 1913 that the first world census of a bird species was published. But already in 1898 Jones (1898) in the state of Ohio had marked in his home town the location of each bird on a map and had determined the number of wintering birds present within three-quarters of a mile of its centre. In 1899, 1900 and 1901 Burns (1901) made a breeding bird census of a square mile of Pennsylvania involving various habitats, and he repeated the census in 1914.

Much of the pioneer bird census work in the United States, which certainly seems to be the original home of modern bird population work, was done in Illinois by S. A. Forbes (1907, 1913) and Forbes and Gross (1921–23). In Britain C. J. and H. G. Alexander (1909) came to the conclusion, after some years' observations of the birds in the neighbourhood of Tunbridge Wells, "that in many species each pair inhabits a definite area into which other pairs do not intrude". With this unconscious acceptance of bird territory as the fundamental census unit, an assumption which lies behind all breeding season census work, the Alexanders began to mark territories on six-inch maps in the spring of 1907, and in that season and 1908 had mapped a considerable area. They discovered comparisons between the breeding populations of seventeen species, of which all but two were passerines, and between these populations in one year and in another.

The birth, then, of the comparative census of various members of a community of breeding birds took place in the early years of the present century, first in the States and shortly afterwards in Britain. Of course, partial censuses of individual species had been performed by observer naturalists for many years before that. However, even after the Renaissance, estimates of bird numbers were mostly vague, even at places where wild birds were exploited and it was presumably therefore useful to know their numbers.

Thus, the breeding gannet population of the Bass Rock was recorded as a "murmuring multitude" in 1521, an "incredible number" in 1526, it "could not be easily estimated" in 1535, there were "plenty" in 1555, in 1570 the "numbers obscure the sun", in 1578 the gannets were "more abundant than Ailsa Craig" (and in the same year the gannet was stated to "abound" at Ailsa Craig). In 1635 the Bass gannets were present in "abundance", in 1661 "in multitudes innumerable", in 1769 still "multitudes". It was not till 1831 that anybody ever put a number to the gannets of Bass Rock, and appropriately enough it was that great and neglected pioneer ornithologist William MacGillivray (Audubon, 1838). MacGillivray found 300 nests on the gravel slope near the landing-place, and wrote that the number of individual birds "might be estimated at 20,000".

Actually, gannet figures go back to the year 1697, in which Martin Martin (1698) made his famous visit to St. Kilda. But he recorded the numbers that the St. Kildans ate, not the population from which they took their crop! The first census ever made of the occupied nests at a gannetry appears to have been that of Friedrich Faber (1822) who found thirteen to fifteen occupied on Grímsey, off the north of Iceland, in 1819, and about twenty nests in the following year.

On 27th and 28th January, 1769 Thomas Pennant visited the famous heronry at Cressy Hall at Surfleet, near Spalding in Lincolnshire, and recorded the presence of over eighty nests in all, a rarity which, commented Gilbert White, "I would ride half as many miles to have a sight of". It was of course because it was something out of the ordinary that Pennant counted the nest population. If it had been an ordinary population he would not have thought to count it. It took White himself to make the first observation of what could be called an 'ordinary' population.

On 13th May, 1778 White (1789) wrote to Daines Barrington:

"Among the many singularities attending those amusing birds, the swifts, I am now confirmed in the opinion that we have every year the same number of pairs invariably; at least the result of my inquiry has been exactly the same for a long time past. The swallows and martins are so numerous, and so widely distributed over the village, that it is hardly possible to recount them; while the swifts, though they do not build in the church, yet so frequently haunt it, and play and rendezvous round it, that they are easily enumerated. The number that I constantly find are eight pairs".

What is so interesting about White's observation is the fact, as we now know, that of all British land-birds, the swift probably has the least mortality in the nest: it is a slow breeding species with a very stable population, just as White says. Incidentally, when W. H. Hudson visited the village in 1896 he found six pairs of swifts. This census of White's appears to be the first local census of a species. The first local census of a community of species appears to have been made by the pioneer American ornithologist Alexander Wilson (1811) who counted fifty-one pairs of breeding birds on eight acres.

Useful summaries of the rise of bird census work and of the techniques and results of bird population studies have been given by Nicholson (1931), Lack (1937) and Kendeigh (1944). They demonstrate abundantly that the two main approaches, the study of individual species and the study of communities or 'habitats' have gone on together, though in recent years the study of the comparative numbers of members of British bird communities has dropped, probably only temporarily, out of fashion. The British woodland survey of Lack and Venables (1939) confined its findings on population to relative abundance of species, not to absolute density. W. B. Alexander's survey (1932) was the first important breeding census of a mixture of pasture, arable land, hedge and copse typical of English farmland; a follow-up series of censuses was conducted by Chapman (1939), and Campbell (1953) has made a census survey of the effect of industry on a Welsh farmland bird community. Lack's survey (1935) is still the latest collection of sample breeding-censuses of British heaths and moorland. There have been many censuses of parkland and gardens, but no important ones since before the Second World War.

These comparative censuses of breeding land-birds have been used by Nicholson and myself to reach estimates of the total breeding-population of England and Wales, of which the latest (Fisher, 1952b) is about sixty million individuals. About 97½ per cent of this is accounted for by the thirty commonest of the breeding species, of which 76 breed regularly and 96 have bred in the present century.

One of the earliest direct population studies of an individual species was a result of a pioneer exercise in co-operative ornithology organized by E. M. Nicholson in Britain. Nicholson, now Director-General of the Nature Conservancy, has done more than anybody to establish co-operative studies in the British Isles. He was largely instrumental in the creation of a new body, the British Trust for Ornithology, founded in 1933, and designed amongst other things to provide a network of amateur bird watchers, skilled, reliable and capable of getting local field information for central analysis. The first great success of co-operative census work was Nicholson's (1929b) *British Birds* census of heronries. Organized through this admirable periodical, his observer network thoroughly covered England and Wales and discovered every heronry in the two countries (with a few exceptions, whose omission affected the total negligibly). In 1928 they found that about 4,000 nests were occupied in about 300 heronries. Subsequently a good sample of these heronries, often more than half, has been covered by volunteers yearly, so that we have had an index of heron population for the last twenty-five years. From this sampling, conducted by Nicholson till 1939 and after that by W. B. Alexander, it was found that the breeding population remained approximately the same, with about 4,000 nests, till 1939. After the cold winter of 1940 it began to fall, and by 1942 it was probably under 3,000. But in 1945–46 it had risen again to the 4,000 nests total, which appears to be the normal. After the cold winter of 1947 the population fell sharply to not much over 2,000 occupied nests. There was no sign of recovery in 1948, but about 3,000 nests in 1949, and by 1950 it was nearly back to 'normal'. This history of the heron population of

England and Wales shows most clearly the value of scientific census work with comparative measurements. The counting of nests in a heronry can be surprisingly difficult, but the experience of British observers in nest censuses of social birds is now very great. There can now be scarcely a member of an important national or local bird society who has not done some nest-counting in his time.

The nest-counting for the heron index, in complexity, intricacy and organization, is rivalled by the vast amount of census work that has been done on the nesting population of the rook, particularly in Britain but also in France, Holland, Sweden, Germany, and Hungary. As I was the organizer of the wartime rook investigation, during which more rookery census work was done than in all the many previous rookery censuses put together, I must be forgiven if I go into some detail.

The pioneer of rookery census-makers were the Scots J. A. Harvie-Brown (many publications) and George Muirhead (1889). As early as 1875 Harvie-Brown, by correspondence with landlords, had made what he believed to be a complete rookery census of Caithness. In 1878 he added Orkney, where there was one small rookery, the Isle of Arran and Sutherland. In 1879 he covered Iona and Mull, in about 1885 Islay. But the rookery populations of these parts of the Highlands were relatively low, the largest being that of Sutherland with just over 2,000 nests, though Caithness with just over 1,000 nests has nearly twice as many to the square mile. Rookeries in these Highland regions tend to be isolated, and fairly easy to find and count, even though many of the nests are in Scots pines. But Harvie-Brown's industrious correspondence with his friends was small compared with that of Muirhead in the year 1887, when he covered the whole of Berwickshire, which, though less than a quarter of the size of Sutherland, had more than ten times as many nests. In February 1887 Muirhead sent out circulars to correspondents in every parish in the country —landowners, farmers, gamekeepers, gardeners, doctors, ministers, schoolmasters—the people in the best position to know the rookeries. His correspondents only returned the population of the larger rookeries correct to the nearest hundred; so that Muirhead's total of 21,000 nests was only a first approximation. Though it may be no more than a remarkable coincidence, the repeat census of Berwickshire in 1945 by an accurate trained observer, George Edwards, also showed 21,000 nests (to the nearest thousand).

There was no more rookery census work in Britain until 1908, when H. S. Gladstone (1910) made a census of the rookeries of the large county of Dumfries. This indefatigable and accurate worker again conducted his investigations by correspondence with local landowners, tenants, medical practitioners and incumbents. In 1921 Gladstone (1922) repeated the census, recording a slight decrease, and in the same year W. Evans (1922) did a first census of the rookeries of Greater Edinburgh. In 1928 Nash (1928) repeated this census and took in also the rest of Midlothian. In 1922 Stewart (1923) did a census of the whole large county of Lanark, which he repeated in 1929 (1930), showing an increase. The Edinburgh census of Evans in 1921 was the first important census where the observer himself found all the rookeries and attempted accurately to count the precise number of nests, instead of relying on correspondence with local occupiers. Since then, all rookery censuses have been carried out on this principle, with groups of observers methodically working all the roads by motor-car, motor-cycle and bicycle, until not an acre of dead ground has been overlooked. (It is surprising how big a rookery can hide in a small fold of land.) This careful and methodical scouring of the countryside by the observers was the note on which rookery census work began in England, though it must be freely admitted that the Scottish workers were the first in the field and the first to tackle the problem on the right lines. Some early English rookery censuses include fifty square miles of the Harrow area in 1928, with a repeat in 1929 (Harrisson, 1931) and 224 square miles round Oxford, also in 1928 (E. M. and B. D. Nicholson, 1930). This Oxford area has almost become the 'control' area for rookery research, because the wartime rookery investigation took place in it, and repeat censuses were done by W. B. Alexander (1933) in 1931, and by Alexander and others in 1944. To cut a story short, by the late 1930s accurate rookery censuses had been made of no less than 8,384 square miles in England, about half of its total area, in the whole of the Isle of Man and a small part of Wales. The area surveyed included the complete counties of Somerset, the Isle of Wight, Lincolnshire, Rutland, Leicestershire, Nottinghamshire, Derbyshire and most of Oxfordshire. All these areas had repeat censuses done in 1944–46; and in those years the rookery census was spread to cover no less than 61.4 per cent of the entire area of England, the Isle of Man, 29.2 per cent of the area of Wales and 88.7 per cent of the area of Scotland. Nearly a million nests were actually counted in over 60,000 square miles.

During the rook investigation no less than 350 members and correspondents of the British Trust for Ornithology conducted rookery census work in the field, and if the number of helpers that these workers took with them on their rookery hunting be included, the number of workers in the field actually counting nests in the Aprils (the only possible month) of 1944, 1945 and 1946, was about 400. They covered an area of about 150 square miles each in a season, as few of them worked in more than one season. Nicholson (1932) had calculated the rook population of England and Wales as approximately 1,750,000 individuals. This was based on the samples surveyed by himself and others in the late 'twenties, and was an accurate estimate, for the figure resulting from the much larger survey in the mid-'forties was a little over two million, and the population was proved to have increased in the meantime. Thus it is that we know, thanks to its social nesting habits and the energy of British bird-watchers, the actual population of Britain's commonest large bird (or largest common bird).

One land-bird has lent itself to census work by its particular conspicuousness. The stork nested in Edinburgh in the fifteenth century but has never nested in Britain since. But on the continent of Europe it has a wide distribution from central Europe to south Sweden and there is another big group from central and south-west Spain and south Portugal to North Africa. Scientific interest in the stork has always been acute, for it has been the subject not only of census work but also of intensive migration studies through marking; and the literature of this bird is voluminous. Perhaps the most useful recent summaries of our knowledge of the species are those of

Schuz (1936), and Haverschmidt (1949). Nearly twenty years ago, in the mid-'thirties, especially in 1934, there was an almost complete census of the stork in Europe. Storks' nests are very obvious and conspicuous, and almost invariably associated with human habitations. In many parts of Europe, if they do not nest on houses, storks nest on cart-wheels placed on the tops of poles. During the great census every (or practically every) stork's nest was found in Switzerland, France, Holland, Denmark, Sweden, Germany, Danzig, Poland, Memel, Latvia, Czecho-Slovakia, Austria, Spain and French North Africa. Both Germany and French North Africa had over 30,000 nests each.

We can list satisfactory censuses of large parts of the populations of important and interesting birds. Thus, the great crested grebe in England, Wales and Scotland had been reduced, largely by the plume trade, to forty-two pairs in 1860, but increased with protection to about 1,240 pairs in 1931. Harrisson and Hollom (1932–33) who conducted the co-operative census of 1931, made a most satisfactory job of it, judging by the small number of amendments which have had to be made to the figures published in their paper. Every piece of water big enough to hold a pair of grebes in the British Isles is of course well known to bird-watchers, and was so twenty years ago. Hollom has conducted sample index-censuses of the great crested grebe in years since, and has shown that the population has remained fairly constant. A more ambitious attempt at a census was the investigation by Blaker (1933) of the number of occupied nests of the barn-owl in England and Wales in 1932. Unfortunately it is not possible to be certain of finding all barn-owl nests, even though a large number nest in barns, and the accuracy of control possible in the cases of rook, grebe and stork could not be attained. More successful have been censuses of substantial parts of the world population of certain sea-birds by counting occupied nests at colonies (those sea-birds whose world population has been counted or estimated are dealt with in the previous chapter). An early sea-bird survey was that of Harrisson and Hurrell (1933) who, in the years round 1930, estimated that between 600 and 800 great black-back nests were occupied in the

Scilly Islands, about 180 in Pembrokeshire, and perhaps 210 or a little more in the whole of the rest of England and Wales. In 1900 this now increasing species had been practically extinct in England and Wales. Fisher and Lockley (1953) list a substantial catalogue of similar partial censuses of sea-bird species along substantial stretches of coast. Those involving populations of over 1,000 occupied nests include the fulmar in Britain, about 100,000 in 1949 (Fisher, 1952), in West Greenland, 200,000 (Salomonsen, 1950), Cory's shearwater *c.* 20,000 on the Salvages Islands in 1939 (Lockley, 1952), double-crested cormorant, over 10,000 in the eastern United States in 1944 (Gross, 1944), European cormorant in Holland in several different years—4,359 in 1941 (van Ijzendoorn, 1950), the same species in North America, 1,086 in 1940 (Lewis, 1941), great skua in Great Britain, about 1,000 around 1946 (several sources), ring-billed gull in the Gulf of St. Lawrence, *c.* 1,750 in 1940 (Lewis, 1941b), common gull in Germany, 20,221 in 1939 (Schulz, 1947), herring-gull in Holland, *c.* 27,500 in 1938 (van Ijzendoorn, 1950), North Sea coast of Germany, 28,569 in 1939 (Schulz) and the same species in Maine in several different years, culminating with 25,000 in 1931 (Palmer, 1949), great black-back in England and Wales *c.* 1,100 in 1930 (Harrisson and Hurrell), the same species in the United States at least 1,250 in 1944 (Gross, 1945), black-headed gull in England and Wales, estimated 75,000 in 1938 (Hollom, 1940), subsequently modified by Marchant (1952) to *c.* 35,000, owing to a miscalculation of the size of the largest colony, kittiwake in England, 6,000 to 8,000 in late 1940s (Fisher), in Newfoundland, 24,400 in 1941–45 (Peters and Burleigh, 1951), in Gulf of St. Lawrence, *c.* 10,000 in 1940 (Lewis, 1941b), guillemot in Sweden, *c.* 25,000 pairs in early 1940s (Salomonsen, 1944), at least 40,000 pairs in eastern North America (Johnson, 1940).

The terns are the most unpredictable and erratic in their breeding habits of all sea-birds, and I will not weary the reader with many censuses of large colonies or groups of colonies involving more than 1,000 nests. It is enough to say that very adequate censuses have been made of the common tern in Maine, near the important Cape Cod region of Massachusetts, along the whole coast of Germany, in most of the principal sanctuaries of Holland, and in the whole of England and Wales. The arctic tern's numbers are known, or have been known, also in Maine, Germany, Holland, England and Wales. With the exception of one or two colonies, all the nests of the roseate tern in Europe (over 1,000) are in the British Isles. A security blackout on the location of the colonies has made the censuses difficult to collate, but there is no doubt that a reliable count of occupied nests has been made at every one of them in more than one of the last ten years. The population of the Sandwich tern in the 1940s has been known approximately in the whole of north-west Europe, including Holland, Sweden, Germany, Danzig, Britain, Brittany and Portugal. The population of the noddy and sooty terns in the United States has been determined in many years of the present century; they only breed on the Dry Tortugas. D. Lack (1937) was already correct when he said that "censuses of local sea-bird colonies have been frequently made and there is not space to refer to them here".

Passing on to game, we find an interesting contrast between knowledge of population in the United States and in the Old World. I should not be surprised if the principal reason did not prove to be the fact that in America game-birds are not private property but the property of the individual States, and their conservation has been financed for years by the state authorities in the interest of those holding hunting licences. The majority of the United States are able to afford research and administration units for game conservation and control which has resulted in the publication of many admirable papers on the game-bird species by energetic workers; this contrasts almost violently with the scientific work on game-birds in Europe. Some of the books on game in Britain written by sporting amateur naturalists before the First World War have never been bettered, which, while it does credit to the sporting naturalists themselves, is not an ideal situation.

Many of the American game studies have involved censuses of substantial areas. Game-birds are not easy to count; but the valuable papers are too numerous to quote. I may mention work on the bob-white by Errington in Iowa and Wisconsin, with the valuable lessons that can be learned from it about numerical fluctuation and predation. Even such a 'difficult' species to census as the ruffed grouse, a woodland species and the most important upland game-bird of the forested areas of the northern and eastern United States, has been successfully counted over large areas, for instance by King (1937). Other birds studied include the prairie-chicken, the California quail and the introduced Hungarian partridge, and among their many able census-takers are, besides those already mentioned, Stokes, Hamerstrom, Gross, Aldo and Starker Leopold, Stoddard, Sumner and Hawkins. To give an example of the enterprising nature of some of these censuses of difficult and elusive species, it is enough to quote Mosby and Handley (1943) who estimated the whole population of the subspecies of the wild turkey that lives in Florida, *Meleagris gallopavo osceola*, at about 12,300 individuals.

Another comparable census, though of a wildfowl, not a game-bird, was the careful establishing of the origin and habits of the big population of Canada geese that use the Mississipi flyway. These belong to the subspecies *Branta canadensis interior*. Hanson and Smith (1950) calculated that the population of Canada geese belonging to this subspecies wintering in the entire Mississipi valley in January 1948 was about 97,000, and a year later *c*. 160,000. Hanson and Griffith (1952) have also investigated the population of this species on the Atlantic flyway. Yet another wildfowl census which embraces practically the whole world population of a subspecies is the assessment of the Pacific black brant, *Branta bernicla nigricans*, by Leopold and Smith (1953), who found 174,740 wintering in North America.

CHAPTER X

CHANGING BIRD POPULATIONS

ONE of the most exciting books of ideas was Ellsworth Huntington's *Mainsprings of Civilization*, which was published in 1945. Among the many novel and intriguing collections of knowledge in this book are several chapters on cycles, on the apparently rhythmical repetition of 'conditions' in the external world. Huntington, like many other workers, was fascinated by periodicity in nature, and he spent much time going beyond those cycles which are familiar to us all, such as the day, the moon-month and the year, in an effort to identify other cycles longer than a year. Huntington appears to accept the existence of some of these cycles uncritically, and to a somewhat suspicious (and non-statistical) eye some of the graphs he produces seem like special pleading. But first approximations are truly valuable, and never harm those bold enough to state them; and the cycle part of his book is most valuable to the ecologist. The cycles mentioned include (with what follow, or are alleged to follow, them):

2 years: fruit crops.
3·4 years (41 months): business activity and atmospheric electricity.
$3\frac{2}{3}$ years: Nile floods and atmospheric pressure.
4 years: animal population.
$7\frac{1}{2}$ years: tree-growth, lake deposits and atmospheric pressure.
8 years: wheat prices.
9·2 years: prices.
$9\frac{2}{3}$ years: prices, animal population, tree-growth and atmospheric ozone.
11·2 years: sunspots.
22·3 years: sunspots.
35 years: European weather.

45 years: weather.
54 years: commodity prices.
68 years: commodity prices and Nile floods.
90 years: weather.
170 years: drought and civil war.
510 years: drought.

In his classic *Voles, Mice and Lemmings*, Charles Elton (1942) established with a wealth of examples unparalleled in animal population studies, that most of the common rodents of the northern hemisphere, and particularly lemmings, showed a marked four-year cycle in their population. Occasionally a cycle was prolonged or shortened by a year, and often cycles in one part of the world might be out of phase with cycles in another part not very far away. But in general the phenomenal universality of the four-year cycle was beautifully demonstrated in Elton's book, and upon this he built a lucid and interesting discussion on the consequent effects of this fundamental cycle upon the animals which live on rodents. The most important of these are other mammals, particularly foxes and the members of the weasel family. But there are also many birds, such as various species of buzzard, owls, skuas and harriers—notably the rough-legged buzzard, the snowy owl *Nyctea nyctea*, the long-tailed skua *Stercorarius longicaudus*, and the hen-harrier *Circus cyaneus*. (In North America the hen-harrier has its counterpart in *Circus hudsonius*, the marsh-hawk.) Other birds which prey on rodents, and whose European numbers may consequently fluctuate with theirs are the Lapp owl, *Strix lapponica*, the short-eared owl, *Asio otus*, the arctic skua, *Stercorarius parasiticus*, and perhaps the eagle-owl, *Bubo bubo*, though this last eats game-birds as well as rodents, and as game-birds appear also to have periodic fluctuations and as these fluctuations do not necessarily march with those of the rodents, the situation is somewhat complicated. Nevertheless the eagle-owl, and also the goshawk, *Accipiter gentilis*, and the sparrow-hawk, *A. nisus*, likewise show strong cycles in numbers, as was first shown by Johnsen (1929). The fluctuations in the numbers of game-birds, which do not depend on rodents for food, are for that reason all the more interesting. The causes of both rodent and game-bird fluctuations are scarcely known, though we are beginning to understand some of the regulating mechanisms. The pioneer work on game-bird fluctuation in Europe is that of Kloster (1921) who demonstrated a good four-year cycle in the numbers of the willow-grouse or rype, *Lagopus lagopus*, with which our native British red grouse, *L. scoticus*, forms an allopatric species-pair. The rype cycle in Norway has actually marched for many years with that of the lemmings, though this may be purely coincidental. In Scotland, where there are no lemmings, the Scottish red grouse does not fluctuate in harmony with the voles, which show the usual four-year cycle, but has a longer period, which was identified by Leopold and Ball (1931) from shooting bags as about six and a half years, the intervals between peaks being sometimes as little as four years or as many as eight.

Middleton's subsequent work (1934) brought the average interval down to 5·9 years. However, the latest investigation by MacKenzie (1952) of bags of red grouse from some eighty moors (some records going back to 1834) show twenty peaks in 105 years, 1837–1941, an average of 5·3 years with intervals varying from three to ten years. MacKenzie, aided by Moran (1952) has also demonstrated that the other three British tetraonid game-birds show rises and falls almost like those of the grouse. These are the capercaillie, ptarmigan and black grouse. In the numbers of black game, the short fluctuations (for which there are thirteen waves in fifty-eight years, an average of about four and a half years) are superimposed on longer waves of general abundance with a period of some thirty to fifty years. The North American grouse cycle was found by Leopold and Ball (1931) to be ten years or a little more, that is, just about twice the length of the European grouse cycle. In Wisconsin, Schorger (1944) found the cycles of the prairie-chicken and the sharp-tailed grouse, *Pedioecetes phasianellus*, to vary from nine to eleven years, and Criddle (1930) also found a nine to eleven year cycle for the sharp-tailed grouse in Manitoba. The classic North American example of a fluctuating game species is the ruffed grouse, *Bonasa umbellus* (Clarke, 1936). It has a period apparently the same as that of the sharp-tailed grouse. The bob-white (Errington, 1945) appears to follow the general North American grouse cycle trends. It is not known exactly how some of the more northerly New World game-birds fluctuate, but Comeau (1909) has recorded some data for Canadian Labrador which suggests a cycle of nine or ten years for the willow-ptarmigan, *Lagopus lagopus*. This is the same species as the rype of Norway, and it is interesting to note that it appears to show the longer North American cycle rather than the shorter European cycle of the Norwegian race.

To return to Europe, there are very few other bird species in which anything like a cycle has been satisfactorily demonstrated. Middleton has shown a periodicity averaging about eight years in the number of partridges, *Perdix perdix*, in England, a cycle which appears to march with that in the numbers of rabbits and hares. Actually his average was 8·2 years for the partridge, 8·6 for rabbits, 7·5 for hares, so theoretically they should get out of phase. But on some estates examined by Middleton they synchronized very closely. Matheson (1953) found that partridge-bags from a Denbighshire estate over ninety-five years showed a periodic interval of from seven to eleven, average 9·5 years. Middleton also found from the bag on an estate in Sussex an indication that the numbers of the woodcock, *Scolopax rusticola*, had a distinct cyclic fluctuation with an average of nine years. Alexander (1947), who analysed six estates, came to a similar conclusion; he found the periods in which at least four of the six estates made especially good bags (all were in Britain) were at intervals of from six to eleven years, with an average of eight years.

Although the study of cycles is now occupying a large number of research workers in special institutes, university departments and societies, we are not yet able to understand whether the fundamental cause of the cycles in the populations of some animals is environmental or not. It is known that there are periodic cycles in the solar system, of which the most notable are those of the sun's spots. It is not yet known whether these lie in any causal relationship with such cycles as have been detected in the earth's weather. That long-term cycles in the earth's weather affect the numbers and distribution of plants and animals is obvious. But it is still quite unknown whether there is any link between weather cycles and the shorter, more clear-cut animal cycles such as the four-year rodent cycle, the six-year Old World game-bird cycle and the nine-year New World game-bird cycle. If we are unable to find weather cycles to which these rodent and game-bird cycles can be shown to be consequential—and so far we cannot do so—what alternative explanations are available? Perhaps the most fruitful suggestions are those of Gause (1934) and Chitty (1952). Gause's thesis, which has been developed by Lotka (1934) and by Volterra and d'Ancona (1935) has shown on theoretical and experimental grounds what can be expected from the study of the numbers of predators and prey in a closed community. The closed community used by Gause was a test-tube, his predators and prey being different species of protozoa.

Stated broadly, without the mathematical symbols and formulae used by Gause, his conclusions are that the numbers of these organisms change in a perfectly orderly manner, and in a way which can be expressed in a mathematical equation. Let us suppose predators are feeding on a colony of B in an enclosed test-tube. To start with, prey B is in the test-tube alone, living on a plentiful supply of inorganic food. In the absence of A, B will increase in numbers. As soon as A is introduced, however, it will prey on B, with the result that B will begin to decrease in numbers, while A will increase. Since the community is closed (i.e. the test-tube is corked up), this process cannot be interminable, since eventually A might be expected to eat up all B. However, Gause has shown that, providing hiding-places are given, some of B will escape being eaten, and that meanwhile the general decrease of B will cause A, up till then increasing in numbers, to decrease from starvation. Upon the decrease of A, B will again start to increase, until a point is reached when A begins to take toll once more, and the process is reversed to the condition at the beginning of the experiment. It can be seen from this that the period of large numbers of the prey B will be followed by periods of large numbers of predator A, but that there will always be a lag period between the peak numbers of prey and the peak numbers of predators, and the low-water mark of each.

Theoretically, then, a regular cycle can be constructed out of the simple predator-prey relationship, out of the transformation of cause into consequence and consequence into cause. An increase in the numbers of prey has as its consequence the increase of predators. This consequence next becomes a cause, the cause of a decrease in the numbers of prey. This consequence in turn becomes the cause of the decrease in the numbers of the predators, and so the cycle goes on. It needs to be clearly stated that, while this has been shown to work with protozoa in a test-tube, it has not in any way been proved to be the cause of the observed fluctuations of vertebrate predators and prey in nature.

The work of Chitty (1952) was concerned primarily with mortality among field-voles, *Microtus agrestis*, during and after one of the peaks of their population cycles. In a four-year study at Lake Vyrnwy in Wales he found much fighting and a high mortality of young in the breeding season at the population peak: the surviving young, moreover, had an abnormally low expectation of life. J. R. Clarke (1953) also showed experimentally that fighting among voles caused a disturbance of their glandular (adreno-pituitary) system. Chitty thinks that the defects of the young may be due to defects in the hormone balance of their mothers, itself a consequence of the observed intraspecific strife and fighting. This would certainly result in alternation of abundance and scarcity: the problem is:—why are cycles so regular, and so often the same over wide areas? It could be that over these cycles the weather may act as a sort of 'governor', for instance by synchronizing the opportunities of population recovery.

It is quite clear that there is a self-regulating mechanism controlling the population of those vertebrates, the small rodents, which are at the bottom of the pyramid of land vertebrate communities, and the most important staple of the vertebrate predators.

If we can accept this interesting thesis, it can be seen that much else falls into place, for the observed predator cycles become simple consequences of the rodent cycle. But what about the game-bird cycles? Game-birds do not live on rodents. On the other hand, they are predated by some of the predators which also prey on rodents. Peak rodent abundance gives rise to peak predator abundance, but after a lag. Usually this lag gives rise to a predator peak one year, sometimes two, after that of the rodents, and usually it coincides with a situation when the rodent population, as a result of its 'crash', is near its lowest. The predators must therefore starve and die, or turn their attention to alternative prey. The evidence suggests that they do both. Hence it is possible that the game-bird cycles may be the consequence of fluctuating predator pressure which is itself the consequence of rodent cycles. Theoretically this should produce a game-bird cycle which is the same in period as the rodent cycle. This seems to work for the willow-grouse in Norway, which, as we have already seen from the work of Johnsen, follows the rodent cycle closely. (However, this may, of course, be due solely to the governing or unifying influence of the weather.)

Elsewhere, as we have seen, the game-bird cycles at six to seven or nine to ten years appear to bear no relation to rodent cycles. This does not necessarily mean, however, that they can bear no relation to the cycles of predators which are fundamentally dependent on rodents, simply because the rodent fluctuations do not always have by any means the same depth, so that the pressure on the predators to seek food elsewhere may be quite different in successive cycles. But it is of course quite possible that the game-bird cycles may, as those of the rodents, be due wholly or partly to self-regulating mechanisms and have nothing to do with predator pressure. Aggression, display and fighting is especially important in the lives of game-birds, in most of which the males have display grounds, and it is quite possible (though so far unproved) that high density on these display grounds may strain the birds' propensity to display and fight beyond breeding-point.

We cannot leave the subject of fluctuations without considering disease as a regulating factor, a subject which Elton goes into in great detail (1942). At peak populations it is unquestionable that the incidence of disease among both rodents and game-birds becomes abnormally high, and in certain years of particularly high peak may reach epidemic proportions, producing a correspondingly deep crash. The most common diseases in game-birds are coccidiosis and gapes (strongylosis). In his recent analysis of the numbers of British tetraonids

MacKenzie cannot bring himself to say whether strongylosis is one of the primary causes of crashes, or "only a feature which sometimes accompanies and intensifies 'drops' due to other causes". One thing seems probable. The numbers of cycle-fluctuating animals, such as rodents and game-birds, vegetable feeders at the bottom of the vertebrate pyramid, are not controlled *directly* by the amount of food available. This is not to say that the animal populations are not attuned to the carrying capacity of the land, because they seem to be so; only among certain rodents, and that at not every peak, are there marked migrations due to the exhaustion of food. The fact is that the factors that control the numbers of animals are complex, multiple and subtle. The history of the investigation of these factors has been a process of rejecting the obvious only to bring it back again with a new slant.

The rest of this chapter will be devoted to changes in bird populations which have exhibited or which still exhibit positive or negative trends which have transcended any cyclic fluctuation. In advance I will state that I shall succumb freely to the temptation to correlate these changes with the most obvious environmental changes that can be found. It is necessary to work by selecting the most economical hypothesis, even if we know very well that the resulting explanation may be only a first approximation to one aspect of the truth.

Without the rise of amateur ornithology in the civilized world in the twentieth century, it would have been almost impossible to write the rest of this chapter: for the extraordinary detail in which the geographical distribution of birds can now be mapped, is a result of the disposition to write bird notes as an amusing hobby—chiefly among the professional classes. The time is long since past when the network of bird recorders was an unofficial one, and More (1865) could, for instance, embrace *all* that was 'known' at the time about the county distribution of British breeding birds by consulting a dozen published formal papers and corresponding with only ninety-two people, who at the time were just about all who could have had anything to tell him.

These stalwart correspondents of More's and the handful of published faunas that he had to rely on in preparing his picture of all that was known about British bird distribution contrast strangely with the resources of ornithology as they are today. In order to sum up our knowledge of the geographical distribution in Britain of one well-known species only, it is necessary to read copiously for many weeks and to enter into correspondence with a vast network of organizations and individuals. There are over 300 natural history societies now in Britain, many of which have their own publications which are worth consulting; and the number of people who belong to at least one natural history society is several tens of thousands. Well over 100,000 British individuals and well over 250,000 individual North Americans possess at least one bird book. In order to ascertain the history and present status of the breeding distribution of the rook, Britain's commonest large bird, it was necessary for me to read papers and notes requiring over 1,000 citations, and to enlist the aid of 400 capable field workers and indulge in correspondence with many more; and in my investigation of the distribution of the fulmar I found the number of papers and notes, somewhat surprisingly, slightly more, the number of correspondents and field workers slightly less. In Britain one bird book is now published on an average every week, and as far as I know there is only one man who attempts to read all the printed words on British birds. Not even full-time academic ornithologists can keep up to date with all the scientific ornithological literature of the world, but must concentrate on aspects of it that particularly interest them.

So proficient are bird watchers and so dense are they upon the ground, that at least in Western Europe and in the United States information about the breeding ranges of birds is so intricately known that anybody who knows how to collect it (and this has a technique of its own) can prepare maps and reports of a detail and accuracy which would amaze an ornithologist from the last century and often amazes non-ornithologists of the present time. Moreover, even a slight change of range of a species, however rare (in many ways, the rarer the easier), is usually detected promptly by some band of local naturalists and finds its way into some bird journal or annual report almost equally promptly.

A particular feature of the ornithological literature of Western Europe in the last ten years has been the identification and analysis of range changes. It is possible that some of these changes may have been going on for longer than we thought, but have only been recognized since the rise of the ornithological mass-observers. But the general conclusion is, notwithstanding that the density of observers has changed, that during the present century, and particularly in the period 1920–50, there has been a marked northerly and westerly spread of a very large number of species in Europe, which can be correlated with an equally marked amelioration of the late spring and early summer climate. Rather oddly, nobody has so far made a collection of all these range-changes, for the purpose of comparison. But there is a very large amount of information collected together for Greenland, Iceland, Norway and all the countries with Baltic shores, also for Hungary. The reader is referred to the useful general account of climatic fluctuations in the countries bordering the North Atlantic by Ahlmann (1949). Most of the students of the effect of the climatic improvement in particular countries have also analysed at least temperature trends in their papers.

In parts of the Arctic the winter temperatures have improved even more markedly than those of early summer. Salomonsen (1948), analysing the situation in Greenland, shows that in the present century the white-fronted goose, *Anser albifrons*, has considerably increased in the northernmost parts of its breeding territory; and that three new species have been added recently—the long-billed marsh-wren, *Telmatodytes palustris* (1943), the fieldfare, *Turdus pilaris* (it arrived by drift migration, see p. 105, in 1937), and the Canada goose, *Branta canadensis* (1945 onwards). Besides these breeding species Salomonsen shows that there has been an increased number of accidental visitors, and that some species which used previously to leave the country in winter, such as the mallard, *Anas platyrhynchos*, and the long-tailed duck, *Clangula hyemalis*, may not all do so; while others, such as the kittiwake, *Rissa tridactyla*, delay their departure longer. Guðmundsson (1951) shows that in the last fifty or sixty years no less than seven southern species have become quite established as breeding birds in Iceland—starling, short-eared owl, the shoveler, *Anas clypeata*, the tufted duck, *Aythya fuligula*, and the black-headed, herring- and lesser black-backed gulls. Further, two other species have bred occasionally but are not yet permanently established, the swallow and the coot. Others already established have substantially increased their range, particularly the black-tailed godwit, oystercatcher and gadwall. And as with Greenland, many partially migratory species now winter in Iceland in greater numbers than they did—among them the redwing, merlin, wigeon, snipe, redshank and golden plover.

The most recent list of Faeroe birds, that of Williamson (1948) gives as recent breeding arrivals the twite, house-sparrow, lapwing and common gull.

In Scandinavia the advance north and west has concerned, not unnaturally, a far greater list of species, and it is only possible to quote some examples, mainly from the admirable summaries of Kalela (1946, 1946b, 1949, 1950, 1952) and Lundevall (1953). Species which are moving into Scandinavia from the south include the serin (first time in Sweden 1942, Denmark 1948), red-breasted flycatcher (Denmark 1941, Sweden 1944), great reed-warbler, marsh-warbler (Sweden 1934), aquatic warbler, black redstart, hoopoe, lesser kestrel, marsh- and Montagu's harrier, collared turtle-dove, bittern (returned as breeder to Sweden 1948), red-crested pochard (Denmark *c.* 1935), cormorant (Denmark 1938), avocet, black-headed gull and quail. Some of these have come from an easterly direction as well as from the south: others which have come from south-east or east or even from north-east (that is, from the direction of northern Asia) include the golden oriole, which, though also a southerly species, has extended its range into Sweden *via* Finland, the scarlet grosbeak, ortolan, yellow-breasted, rustic and little buntings, red-throated pipit, among the warblers the greenish, arctic, grasshopper-, reed-, Blyth's reed- and icterine, the sprosser, the red-flanked bluetail, bee-eater, roller, red-footed falcon, pallid harrier, spotted eagle, black kite, gadwall, garganey, black-necked grebe, little gull, pochard, tufted duck and black-tailed godwit. Some of these involve re-colonization of Scandinavia or extension of breeding range of species which had already got a foothold there. But some of the birds which have moved in have come all the way from central Russia or Asia. The pallid harrier is one of the birds that depends on rodents, and several of its invasions of the east Baltic mentioned later (p. 159) may be consequences of rodent crashes in its ordinary range. In 1952 the species was found for the first time breeding in Sweden and north Germany.

But the red-flanked bluetail, *Tarsiger cyanurus*, had, before 1949, only been recorded twice in Europe. It was clearly a drift wanderer in Shetland in October 1947 (Bruce, 1948) and what was clearly another drift specimen was taken at Pisa in Italy in November 1879. Its previous distribution was thought to be limited by the Urals, but in 1949 it suddenly extended its distribution into Finland, and during the following years was found in many places in eastern and central Finland.

The spread of the yellow-breasted bunting, *Emberiza aureola*, west into the north Baltic area is worth mentioning, since it has formed the subject of a special investigation by Timoféeff-Ressovsky (1940—see also Kleinschmidt, 1941). But most interesting of all are the spreads west into Scandinavia of the red-breasted flycatcher, *Muscicapa parva*, the greenish warbler and the arctic warbler. The extraordinary thing about these three species, which, as Moreau (1952) points out, is surely symptomatic of their Asiatic origin, is the fact that they winter exclusively in Asia, the first in India and the last two in Indo-China and Malaysia. Since the western Palearctic members of no other species, as far as is known, winter in Asia (whereas, as Moreau shows, many Asiatic species winter in Africa), we can understand that the arrival of these birds in Scandinavia may be the end of a very considerable spread. The red-breasted flycatcher is probably now an annual breeder in south-eastern Sweden, having markedly extended its range west in northern Europe since about 1920. The first Danish nest was found in 1941, the first Swedish in 1944. The spread of the greenish warbler, *Phylloscopus trochiloides*, is described by Välikangas (1951). This has spread markedly of recent years in Finland, and to eastern Sweden. It is probable that the arctic warbler, *Phylloscopus borealis*, may have been overlooked for many years. It is restricted to birch-woods north of the arctic circle in

Scandinavia, and does not begin to sing till very late in June; until it sings it is very liable to be confused in the field with other leaf-warblers. Lundevall points out that many records in the last few years point to an increase in northern Finland since 1900, but apart from reports of nests in 1916 and 1940, not definitely confirmed, the first Swedish breeding record was made in 1952 at Abisko, where a colony of singing males was located by members of the International Ornithological Congress of 1950.

I would make no further point if I quoted in detail other examples of colonizations of the countries of north-western Europe. The reader is referred to such useful summaries as those of Løppenthin (1948), Salomonsen (1948) and Jespersen (1949) on Denmark; Peitzmeier (1951) on north-western Germany; and Keve (1948) and Keve and Udvardy (1951) on Hungary. But the spreads of some individual species are worth describing in a little detail. Thus Schenk (1944) describes the recent spread of the olivaceous warbler, *Hippolais pallida*, into the Hungarian Plain from the Lower Danube; in thirteen years it advanced over 115 miles. The Syrian pied woodpecker, *Dryobates syriacus*, unknown in Europe before 1890, was discovered in the Lower Danube in 1928, and by 1950 had advanced over 160 miles (Keve and Udvardy, 1951). This woodpecker has now reached southern Slovakia and Austria, where it may have bred in 1951 and certainly did so in 1952 (Bauer, 1952; Mächler, 1952). The remarkable spread of the serin through Europe has been studied by Schenk (1926), in great detail by Mayr (1926), also by Rensch (1941) and Delamain (1933). In about 1800 its range extended not much further north than Yugoslavia, Italy and the southern part of France, with an extension up the Rhône into the Upper Rhineland and another up the Danube into the western part of the Carpathian basin, where it was first seen in 1834. Rather slowly, but quite surely, a spread subsequently took place at first further up the Danube and down the Rhineland and into the German plain, but also west through France.

> "The left wing", writes Delamain, in France, "advances as far as the Loire, crossing Limousin, Saintonge and Angoumois. In 1903 the Serin reaches the edge of Paris

> which he actually conquers, as well as Champagne, in 1906. During the Great War, in 1915, he sings gaily in bombarded Soissons. Then he pushes on as far as Belgium, and the line of the front forms a salient which is partly reduced by the occupation of Picardy. In 1929 he reaches the furthest point of Finisterre, and he who loves the heat of African oases is so much at home on the Gulf-Stream-washed shores of Brittany that he establishes a colony there. Finally 1931 finds him in Deauville. He is now master of the French side of the Channel.
>
> How far will the Serin go? Probably he covets England where at present he is seldom found."

The serin has not yet been proved to breed on the British side of the Channel but it is still extending its range in Europe north-westwards; it possibly bred in Sweden in 1938 and became a regular breeder in south Sweden in 1942; the first Danish nest was found in 1948 (Lundevall, 1953).

But of all the north-westerly spreads across Europe, that of the collared turtle-dove, *Streptopelia decaocto*, has been the most remarkable (Fisher, 1953). This is a species which may have spread originally from India. It is a member (p. 102) of a superspecies which also includes *S. roseogrisea* of Africa. The details of its spread can only be picked up in the seventeenth or eighteenth centuries, when it seems that it reached Constantinople, and by the early years of the present century we find it established in an European outpost from southern Yugoslavia and Albania across to southern Bulgaria and northern Greece. In 1912 the species arrived at Belgrade, but there was no further sign of spread up the Danube till about 1932, when it was recorded at Monor in Hungary. At this point the species seems to have suddenly become dynamic, for in 20 years—1932–52—it reached 14 new countries, had bred in 11 of them, and had been recorded in the literature from nearly 500 different places. 1,200 miles in 20 years represents a remarkable extension of range, but it bred in Sweden in 1952 and in the same year a solitary male occupied a territory in Lincolnshire, for four months (May and Fisher, 1953); it may have belonged to the spread; and what was probably the same bird returned to the same territory in 1953. The spread seems

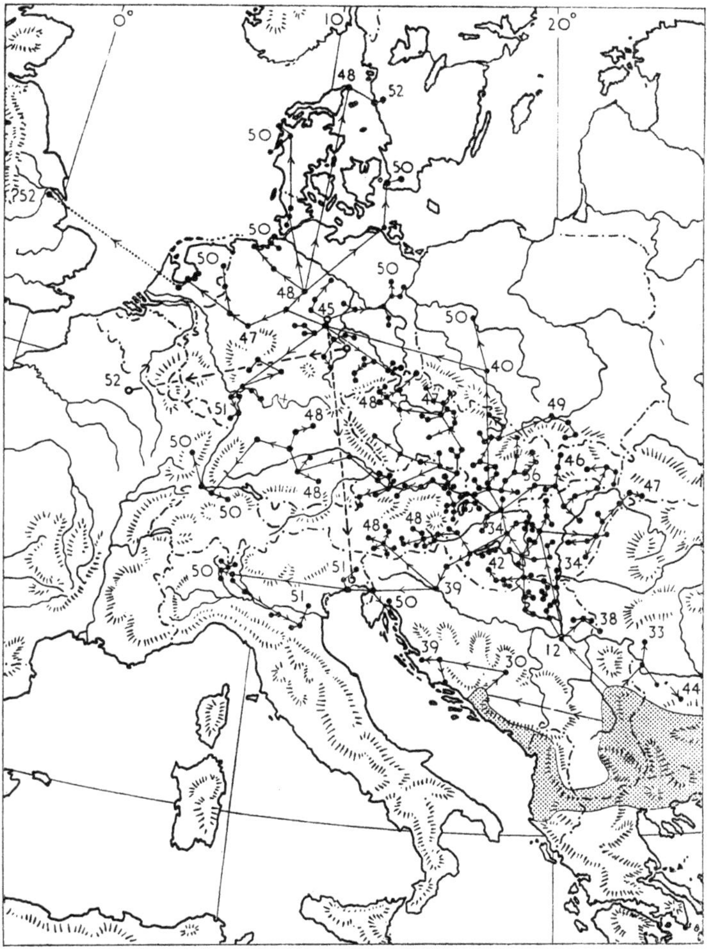

Fig. 2. *The collared turtle-dove in Europe, from Fisher* (1953). *Shaded area represents that occupied by* c. 1900. *Subsequent dates shown concern important places only. Unbroken lines connect localities with nearest places at which previously recorded. Broken lines connect origin- and recovery-spots of two ringing returns.*

too powerful to be attributed to the ameliorating climate, and may have something to do with its attachments to human settlements. Nicholson (1952) suggests that it is possible that both climatic and human factors are giving assistance.

So far we have not described decreases of population which are apparent consequences of the climatic amelioration. These do not feature in the literature to any great extent, and certainly involve many fewer species, most of them arctic and sub-arctic birds. Of quite a number of these there has been a noticeable northward withdrawal of the southern limits of the breeding range. Thus in Finland the broad belt of overlap between the breeding ranges of the chaffinch and the brambling is bodily moving northwards. The arctic tern is becoming a less common breeder in England, which is at the southward limit of its range, and the common tern a more common breeder in Scotland, the Hebrides and the Northern Isles, which are at the northward limit of its range.

Perhaps the most arctic of the sea ducks is the long-tailed duck or old-squaw, which breeds all round the shores and tundras bordering the Polar Basin, the southernmost extensions of its breeding range in the North Atlantic area being in Iceland, where it is an important element of the duck population, the Faeroes, Orkney and Shetland. From Iceland to Orkney there is every indication of a recession in its breeding population. It seems clear that between 1831 and 1926, in both of which years there were probable, though not quite provable breeding records, the long-tailed duck nested irregularly in Orkney. It also seems clear, though the evidence must be regarded as circumstantial, that the long-tailed duck occasionally bred in Shetland, at least between 1848 and 1887. In the Faeroes its presence as a breeding species has always been rather uncertain. But from all three groups[1] there has been no suspicion of a breeding record whatever since the uncertain Orkney record of 1911 and the probable Orkney records of 1912 and 1926, though there is always a small summer population, probably not fully mature.

[1]In those northern islands there is also a regular summer non-breeding population of turnstones, *Arenaria interpres*, great northern divers, *Gavia immer*, velvet-scoters, *Melanitta fusca*, and other species.

In Iceland the long-tailed duck bred abundantly in the coastal areas and particularly on the astonishing eutrophic (fertile) lake of Mývatn in the north. Here it was once the second commonest species of duck. Since the beginning of the century, however, it has greatly decreased at Mývatn (Guðmundsson, 1951) and has to a great extent been replaced by the tufted duck, a new immigrant. It has also decreased in the coastal marshes but is one of the few species that is found nesting in the vegetation oases in the tundra of the 2,000 foot central Iceland plateau (Scott, Fisher and Guðmundsson, 1953). It is possible that a part of the Icelandic population is retreating elsewhere, as five long-tailed ducks ringed at Mývatn have been rather recently recorded in south-west Greenland.

The little auk, *Plautus alle*, has an even more high arctic breeding distribution than the long-tailed duck. Its southerly outpost is in north Iceland. Formerly, according to Guðmundsson, this consisted of two stations, one on the island of Grímsey, just within the Arctic Circle, the other on Langanes, the extreme tip of the most north-easterly promontory of Iceland. The latter, which does not appear widely in the literature, was occupied by a small colony until 1928 or 1929, when it became deserted after a steady decrease during the previous years of the century. The well-known colony at Grímsey is now also nearly extinct, and has evidently never been large during the last 130 years. In 1903 Bernhard Hantzsch (1905), a pioneer ornithographer of Iceland, found 150–200 pairs on the island in two groups. In 1910 there were only six birds at one of these stations, which was vacated some time before 1933. In 1933 Lack and Roberts (1934) found only five birds at the other, but in 1934 Holmes and Keith (1936) saw up to twenty at a time. In 1949 separate visits by Guðmundsson (1951) and Foster, Baxter and Ball (1951) showed that there were at least nineteen birds but probably not very many more. Thus, since 1903, the population seems to have been reduced from at least 150 pairs to perhaps not more than ten.

The little auk is a bird of the ice-front. Its primary food is plankton, and particularly the large shrimp-like animals *Mysis* and *Thysanoessa*, which are particularly common at the edge of the sea ice. During the recent climatic amelioration, and particularly in the second quarter of the present century, the ice-front has retreated north-west from Iceland, and seldom reaches it even in winter. Probably the distance from Grímsey to the ice-front is already inconvenient for the remaining Iceland little auks, and if the present climatic trend continues the colony will disappear when its present inhabitants, which are presumably tied to it by strong traditional bonds, die off.

So far we have described changes in bird populations and distribution which have probably been due to changes in world weather in the long-term sense, or at least not in the sense of the weather of particular years. As we have seen, we have been unable to link the observed cycles in the numbers of certain birds with short-term weather cycles. But there is no doubt that the abnormality of particular summers or winters can have a profound, and occasionally cataclysmic, effect on bird populations. As far as we can see, departures from the 'norm' due to special seasonal weather are usually downward, by which I mean that we encounter in the literature more comments on the incidence of 'bad' seasons than on the incidence of 'good' seasons. Possibly this is because the human observer tends to regard a good season unconsciously as an average one.

Most of the readers of this book will live in south-east England, north-west Europe, the New England states of North America, or California. These areas, as Huntington (1940) has ingeniously shown, enjoy the greatest reliability of crops in the northern hemisphere, perhaps in the world. In these areas the climate, though subject to the usual long-term trends, is more stable and predictable than elsewhere. Farmers enjoy a more regular income, food prices are most stable. It is difficult for an observer who lives in any of these areas to understand the unreliability and unpredictability of weather in Portugal, the central Balkans, arctic Scandinavia or the northern part of the Middle West, far less the Arctic proper. This difference is reflected by the surprise with which the ecological ornithologists of the '20s and '30s encountered the phenomenon of 'non-breeding' in the Arctic, and the various ingenious suggestions of periodicity that have been made. The situation has been summarized by Fisher and Lockley (1954), who show that it is now a commonplace of arctic observation that after late springs and in hard weather some, or even all, of the local bird populations fail to breed. "In our opinion," they write, "these non-breeding years in the Arctic are simply due to the bad weather closing down the food supply." Thus, in very bad springs, the ice never breaks up in Baffin's Bay and the sea-birds of the bird cliffs of eastern Baffin Island cannot find food within operational range of their breeding colonies. Similar conditions obtain in some years on the east coast of Greenland, particularly from Scoresby Sound northwards, and at least early in the present century in Spitsbergen, though the climatic amelioration seems now to have made it 'more reliable'. In King Charles Land, east of Spitsbergen, only nine species of sea- and shore-birds were present in 1889 when the ice conditions were bad, and twenty-one in 1898 when they were good.

These bad summers affect the bird populations not so much by killing the adults as by inhibiting their reproduction. The inhibition may be total, the birds arriving to find their breeding grounds closed, in which case, after a period of balance, their gonads regress and they go out of breeding condition into moult without fertilizing or laying eggs; conditions of partial infertility may result, when only part of the population attempts to breed, and then with reduced clutch-size; or the clutches may be normal, but the eggs, and more often the young, die through inclemency of weather or lack of food. It is clear that the environment controls reproductive output both indirectly, by its effect upon the birds' receptors and glands, and directly, by killing their eggs or young. The latter process is much more readily observable than the former, though it may not be as important.

Hutchinson (1950, pp. 49–65) summarizes the history of the changes in the birds of the important guano islands of western South America (see also p. 115). On this coast it is well known that catastrophes occasionally overtake the huge bird population.

> "The adults", writes Hutchinson, "may suddenly desert their breeding grounds, leaving thousands of eggs unincubated or chicks dying of hunger. Vast numbers of these

adults may fall dead into the sea, others may [leave Peru and] appear in great flocks in Chile or even Ecuador, and, though part of these irruptive populations may ultimately return to the islands, many perish outside their normal range. Though these occurrences are popularly attributed to a pandemic *peste*, there is impressive evidence first marshalled by Lavalle (1917) for such an event in 1917, that the primary factor including the catastrophe is an abnormal extension southward and coastward of warm water."

This, in extreme cases, constitutes a well-developed current counter to the famous cold northward-flowing Humboldt Current upon which the plankton and fish, and thus the vast sea-bird populations, depend. Hutchinson lists catastrophes also in 1911–12, 1925 1932 and 1939. He demonstrates in admittedly rather a confused situation a tendency towards cycles combining high rainfall with sea-current changes, bird catastrophes and consequently guano output. The intervals between peaks of guano output since that in 1910 were 7, 7, 5, 6 and 5 years. So the guano birds of South America may be proved to enjoy a population cycle which *is* connected with weather.

It is of course a commonplace of garden bird-watching that the clutches and broods of early nesters, such as the blackbird and song-thrush in Britain, are particularly vulnerable to March and April rainstorms and frosts. The results of the Hatching and Fledging Enquiry of the British Trust for Ornithology show a very great variation from one season to another in the success of early nests of these common species. Bad winters can affect bird populations as fundamentally as bad summers. Hard winters in which there was much destruction of birds have been recorded in the literature for 671, 1111, 1115, 1124, 1335, 1407–08, 1462, 1609, 1708, 1716, 1739–40, 1768, 1776, 1784, 1788–89, 1794–95, 1813–14, 1838, 1855, 1860–61, 1878–79, 1880–81, 1890–91, 1895, 1917, 1929, 1939–40 and 1946–47.

In 1407 (Elton, 1927) thrushes and blackbirds and many thousands of smaller birds died of hunger and cold in Britain. In 1716 (G. Fleming, 1871) many goldfinches and other species were destroyed. The famous frosts of January 1776 and 1768 were most felicitously described by Gilbert White. In 1768 redwings and other thrushes were killed by the frost and in 1776 the birds

> "began to be in a very pitiable and starving condition. Tamed by the season, skylarks settled in the streets of towns because they saw the ground was bare; rooks frequented dunghills close to houses; and crows watched horses as they passed and greedily devoured what dropped from them. . . . The thrushes and blackbirds were mostly destroyed; and the partridges by the weather and poachers were so thinned that few remained to breed the following year".

Jourdain and Witherby (1918) show that the winter of 1838 was fatal to many birds, and the terrible cold spell from January to March 1855 (the 'Crimean winter') killed not only many residents but also some of the earlier summer visitors. The winter of 1880–81, one of the most intensely cold on record, also killed large numbers of birds. The frost from 25th November, 1890 to 22nd January, 1891 lasted continuously for fifty-nine days, and its effect on wild life is described by Preston (1891). In Norfolk at that time, according to Southwell, kingfishers, herons and bitterns died of starvation; the mortality among thrushes was very great; fieldfares and redwings appear to have moved out of the country altogether. Even the gulls, which benefit to a certain extent from the deaths of others, suffered severely.

In many hard winters in England conditions do not extend to the West Country (Devon and Cornwall) and the south of Ireland, and these become a refuge for very large numbers of the more mobile species. A cold period started in November 1916 that persisted without any real break until April, which was marked by abnormal and persistent snowfalls, followed by drought, which proved the last straw as far as birds were concerned, for fearful conditions then extended to these last western refuges. Jourdain and Witherby (1918) produced a report of destruction which was much more appalling than anything published before. Birds which could scavenge off the bodies of starved members of small species were not seriously affected, but all native species of finches suffered severely, there was a heavy mortality of grey wagtails and tree-creepers. It was estimated that from 40 to 90 per cent of the great tit population was destroyed, and the long-tailed tit, as is usual in hard winters, was almost exterminated. (This species can recover surprisingly quickly from a 90 per cent destruction.) The bearded tit population of Norfolk was reduced to a few pairs and there was a marked diminution of another extremely local species in Britain, the Dartford warbler. All the native thrushes were hard hit but quickly recovered. Stonechats, robins and wrens were seriously—though variably in different regions—decreased. Hedge-sparrows did better than robins. And the kingfisher seems to have been entirely exterminated in the upper reaches of the Thames. The larger birds did better, and the herons, which are usually seriously hit by a hard winter, seem to have maintained their numbers well. The stock of coots on one of the Norfolk Broads was reduced to twenty-five per cent of its previous population.

The severe weather of February and March 1929 and its effect on bird life is described by Witherby and Jourdain (1929). Although very low temperatures were recorded, especially between 11th and 17th February, conditions were "not on the whole very inimical to bird-life". The West Country and southern Ireland refuge remained open. There were many weather movements to the west. The thrush family suffered particularly. Long-tailed tits were greatly reduced. But the larger birds were affected seriously only in the eastern parts of England, from which many moved to the west. After the cold spell in mid-February large numbers of birds of all kinds, but particularly starlings, redwings, song-thrushes and blackbirds, came into Pembrokeshire and continued west. R. M. Lockley recorded that

> "birds were noted dropping into the sea on the very short passage between the mainland and Skomer and on the very small beach were 40 starlings, some song-thrushes and a blackbird dead, washed up on the tide-line. On Skokholm a conservative estimate of the numbers of the chief victims

calculated from the remains found was as follows: golden plover 240, lapwing 30, starling and redwing 200 each, and song-thrush 100.

The severe weather of 1939–40 is described by Ticehurst, Witherby and Hawke (1940). The general effects in 1940 were far more serious than in 1929, though not so disastrous as 1917. The cold, though severe, was not so prolonged. The finches were hard hit, though not all so badly as in 1917. Wood-larks suffered severely, as did grey wagtails. The tits did not suffer such severe losses as in 1917. The long-tailed tit, which was almost exterminated in 1917 and suffered badly in 1929, decreased also very considerably in 1940, and the bearded tit was also reduced, as was the Dartford warbler—indeed, it was practically exterminated. For the rest, the situation was very much the same as in 1917, except that the kingfisher and the green woodpecker suffered more severely; the heron suffered very severely indeed. Many waders were reduced or driven to make extensive weather movements.

The severe winter of 1946–47 was examined by Ticehurst and Hartley (1948). The bad conditions spread to the West Country. The situation was much the same as in 1940. Nuthatches, which do not always suffer in these hard winters, were hard hit, but the long-tailed tits were not affected quite so generally as in 1940. Once more the bearded tit was practically exterminated in Norfolk, and the Dartford warbler in southern England. As usual kingfishers and green woodpeckers suffered severely; the latter species lives very largely on ants in winter and cannot get at the ant-hills if the ground is frozen hard. The heron population was reduced to the lowest level recorded since records of numbers began in 1928 (p. 125).

Using Nicholson and Alexander's index of heron populations, Lack (1946) shows that the English heron population marches to a certain extent with the mean temperature of the coldest winter month. He points out that one of the features of the heron's population is that after 'crashes' in hard winters there is an extraordinarily quick return to power and thenceforth remarkable stability till the next hard winter. This probably means that in the year after the crash, when the heron density is low, more than the average number of herons survive. Lack thinks this is probably because there is more food to go round and that the normal stable heron population is 'regulated' to the available food.

One feature of the cold weather of early 1947 was that its effect on bird life was investigated not only in England, as here recounted, but also in Scotland (Dacker, 1948) and Ireland (D. Evans, 1948). They reported conditions similar to those in England, though the waders were particularly badly hit both in Scotland and in Ireland, especially curlews and lapwings. There was a large invasion of waxwings which reached Ireland.

Generally the history of these cataclysms is a terrible one at the time; but the recovery of most species is very swift, even with slow-breeding birds, and populations may be back to normal in two years. Even the apparently vulnerable species with small populations, and those on the edge of their range, such as the bearded tit and Dartford warbler, show extraordinary powers of recovery, though the winter of 1946–47 seems to have wiped out the entire British bearded tit population except for one male, our country probably being colonized again from Holland in 1948.

Short periods of abnormal weather may have cataclysmic effects upon birds, though these are naturally not so far-reaching as the effects of bad seasons. Birds are vulnerable to storms, hurricanes and fog. We are only beginning to realize just what winds can do by dispersing birds, altering their populations and introducing them as breeding stock into new parts of the world (see p. 105). Occasionally winds may produce cataclysms among offshore and oceanic sea-birds, catching large populations on lee shores, separating them from their sources of food supply, depriving them of room to manœuvre, and ultimately driving them, starved and exhausted, inland to perish. In England and North America, though not to any extent in Western Europe, hurricanes of great breadth and force may even drive oceanic birds miles inland when these are strong and well-fuelled. Many examples of the effects of hurricanes are quoted by Murphy (1936). Some of the sea-bird cataclysms seem intermittent, in such a way as to give a superficial impression of cyclical trends. But when the crashes in the North

Atlantic of such birds as little auks, Brünnich's guillemots, razorbills, Leach's petrels and storm-petrels are analyzed (some are collected by Fisher and Lockley, 1954), it can be seen that the crashes are quite irregular. Important crashes of little auks on the east side of the Atlantic, mostly in Britain, took place in 1841–42, 1848–49, 1861–62, 1878–79, 1894–95, 1899–1900, 1928–29, 1948–49 and 1949–50. Notable wrecks on the eastern seaboard of the United States took place in 1871–72, 1787–79, 1891–92, 1901–02, 1902–03, 1913–14, 1932–33, 1950–51. The most spectacular recent wreck of Leach's petrels was that of October–November 1952, when some thousands were driven inland and destroyed over most of the British Isles. As many Leach's petrels died as could be accounted for, it would appear, by the total British breeding population, and it seems unquestionable that a large proportion of them must have come across the Atlantic in a north-westerly gale which was the consequence of a deep and persistent area of low pressure south-west of Iceland, from south Labrador, Newfoundland or Maine, where Leach's petrel nests in greater numbers than it does in its seven known colonies on the east side of the Atlantic.

Some of the sudden visits of outlandish species to Britain are not the result of wind conditions or drift, though they can be described as weather-movements. The invasions of the British Isles by waxwings from Scandinavia appear to be getting a little more frequent, and are undoubtedly connected with hard conditions in southern Norway and Sweden, where they normally spend the winter. The visitations seem to be truly irregular, and no stretch of the imagination can see a cycle in them. The following list of waxwing invasions years does not set out to be complete but mentions only those which were most commented upon in the literature: January 1680, March 1686, winters of 1702–03, 1787–88, 1828–29, 1834–45, 1849–50, 1863–64, 1666–67, 1872–73, 1892–93, 1903–04, 1913–14, 1921–22, 1931–32, 1932–33, 1936–37, 1941–42, 1943–44, 1945–46, 1946–47, 1948–49 (small), 1949–50. The increase in recent records in not wholly due to the increase in the number of observers. The last important invasion was described by J. and C. Gibb (1951). The first bird arrived in Britain on 28th October, 1949. A little under 2,000 detectable individuals were eventually recognized by the network of bird-watchers, which compares with about 12,500 in the great invasion of 1946–47 and about 200 in the little one of 1948–49.

Largely because of the beauty of the waxwing and its habit of conspicuously feeding on berries in hedges, its irregular visitations to Britain have been chronicled very extensively. There are many less conspicuous species, such as the brambling, whose winter population in Britain are almost as variable, though there is never a winter when bramblings are absent from the eastern part of England and Scotland. The distribution and movements of this species depend to a certain extent on the beech-mast crop. Other birds which irregularly visit Britain are Pallas's sand-grouse, the rose-coloured pastor or rosy starling, the continental crossbill, the mealy redpoll and the nutcracker. Pallas's sand-grouse, *Syrrhaptes paradoxus*, which normally breeds only as far west as the extreme south-eastern corner of Russia in Europe, performed, until the first decade of the present century, irregular invasions which occasionally took it right across Europe to Britain and even to the Faeroe Islands. Its range extends eastwards across Asia, and it also shows irruptive tendencies in north-eastern Asia. On some of its irruptions in Western Europe it has nested, as in Denmark and as in Yorkshire in 1888 and the Moray Firth in 1888 and 1889. The principal invasions in Britain were in 1863 and 1888, when (Witherby and others, 1940) it arrived along the whole east coast of Great Britain and spread west as far as Donegal and the Hebrides. But birds were recorded in Britain in 1859, 1863, 1872, 1876, 1880, 1888–91, 1899, 1904, 1906, and 1909. What caused the irruptions and why they have stopped are things we do not know. Pallas's sand-grouse is normally a social desert-nesting species.

The invasions of the crossbill and the nutcracker may be consequences of bad weather conditions on the Continent, or the failure of the conifer seed-crops, or both.[1] The invasions of the continental crossbill, *Loxia c. curvirostra*, which as far as can be detected have seldom linked up with the indigenous

[1]My friend Alec Robertson points out that this widely-given explanation cannot fit the great crossbill invasion of Britain in 1953, which is more likely to have been a consequence of a failure of the late spring caterpillar-crop.

group of Scottish crossbills, *L. c. scotica*, in the eastern Highlands, have been described by Fisher (1940, p. 77). The earliest recorded crossbill invasion, incidentally, was in 1251, when Matthew Paris, a monk of St. Albans (Gurney, 1921, p. 59) writes

> "In the course of this year about the fruit-season there appeared, in the orchards chiefly, some remarkable birds which had never before been seen in England, somewhat larger than Larks, which ate the kernel of the fruit and nothing else, whereby the trees were fruitless, to the loss of many. The beaks of these birds were crossed. . . ."

In the years 1838–40, 1899, 1910, 1927–30, 1936 and 1953 these irruptions of continental crossbills have been very large, and after several of the irregular invasions, birds have stayed to breed. Gradually the number of counties in England in which there is a stable breeding population of breeding continental crossbills has increased, and for most of the present century it has been seven or eight; in some particularly successful years, such as 1910, it has nested in no less than eighteen counties. There is an excellent analysis by Witherby and others (1938, vol. 1, pp. 95–96).

As mentioned elsewhere (p. 144), some irruptions may be consequences of the rodent and game-bird cycles. The visitations west in Europe of pallid harriers may take place after crashes in the rodent populations in their normal range, and those of snowy owls southward from nearly all the Arctic regions are known to be the consequence of rodent crashes. In some years snowy owls reach Britain, particularly Shetland.

It is perhaps appropriate to mention in this chapter occasions where cataclysms of weather may operate upon bird populations by suddenly altering their habitat rather than by any direct impact upon the birds' life or health. There are few such cataclysms to be recorded and they are highly irregular. Perhaps the most important of them is flood; also important, though on a comparatively minor scale, are the avalanche and landslide, and in certain limited parts of the world, the phenomena associated with volcanic action—earthquakes, eruptions, lava-flows, ash-falls and the appearance and disappearance of islands.

The final end of the great auk in Iceland was probably hastened by the disappearance under the sea of one of its few nesting skerries in 1830, during the volcanic upheavals of that year. In Iceland, too, cliff falls have destroyed a large proportion of the nests in the small, now extinct, gannet colony in Grímsey, which was the northernmost in the world. Avalanches may destroy many acres of woodland, and thus disperse their attendant bird community, to replace them with recolonizing herbage, with a different bird community. Forest fires may strike over wide areas, and the birds of climax forest give way for some years to the birds of secondary growth. But it takes a prodigious forest fire to destroy a whole woodland community, and in the most severe outbreaks relict stands of trees are nearly always left here and there, from which the birds can recolonize. Averil Morley (1940) studied the bird life of 200 acres of burnt woodland in east Sussex in the first and fourth summer after burning. In the first breeding season only mistle-thrush and nightjar were present. In the fourth six species were present—tree-pipit, willow-warbler, robin, chaffinch, whitethroat and wood-pigeon. Morley identified five 'habitats' in the first season and five in the fourth, to be compared with twelve main habitats in a control piece of unburnt forest, which contained no less than thirty-six species. The recolonization was thus a slow business.

The only cataclysm which can overwhelm vast areas, at least in Britain, appears to be a flood. This applies particularly to the east coast of England from Lincolnshire to Kent, where, on rare occasions at irregular intervals, the sea walls may be broken and almost the entire chain of low drained arable lands, broadlands and coastal marshes may be inundated. On occasions, as in February 1953, almost every marsh in East Anglia has become suddenly covered by the sea. These inundations (Manley, 1952) arise from tidal surges in the southern North Sea, when exceptionally high tides coincide with big depressions west of Norway, and the consequent north-westerly gales. Floods resulting from breaches have devastated eastern coastlands in 1287, 1362 and several other occasions in the fourteenth century, 1608, 1609, 1665, 1703, 1791, 1802, 1805, 1897, 1928, 1938 and 1953. This is not necessarily a complete list, and there is some evidence (Steers, 1953) that many of the past and present meres and salt-marshes of eastern England were caused by actual invasions of the fenlands by sea floods.

A beautiful collection of evidence on the consequences of one of these floods and their effect upon bird populations has been gathered by Buxton (1940–44; see also Sainty and others, 1939). From 10th to 12th February, 1938 Horsey Mere and Hickling Broad in Norfolk were inundated through a large breach in the sea wall. About 7,500 acres were flooded with salt water. The birds divided themselves into three groups—those whose food, when the floods receded (which they did fairly rapidly), was not materially affected, those whose habitat or food was destroyed and which consequently emigrated, and those whose habitat and food were destroyed and which consequently starved. To the last category at least some of the local pheasant and kestrel population appear to have belonged; the pheasants because the fields were flooded, the kestrels presumably because all the mice were drowned. Of those which emigrated (the vast majority) there was a return in almost every case to the marshes when conditions (sometimes after the lapse of many years) once more suited them. It was as if there was a reserve population always ready to exploit the place in nature the moment it reappeared. After the Hickling *débâcle* Montagu's harriers returned after three years, garganeys after one, some bitterns immediately and all after two years; the bearded tits (which are entirely confined to the reeds) decreased but did not vanish, only to be wiped out by a subsequent hard winter, to which they were much more sensitive.

But with the invasion of an abundance of salt-water shrimps there were very many more spotted crakes than usual, feeding on them. Moreover, the inundated areas, after the water had receded, were cleared of over 100 species of salt-sensitive plants, and colonized by a succession of salt-loving ones. These provided better food than their predecessors for some surface-feeding ducks, whose population increased. But, presumably because of the wiping out of the fresh-water fish by the inundation, no diving ducks used the place for some years. There was a marked decrease of coots and moorhens, a marked increase of yellow wagtails.

It can be seen, then, that the consequences of a flooding cataclysm on bird life may be profound, and of course some habitats may be permanently changed by flooding. It was flooding, incidentally, that brought back for a short period during the war the black tern as a breeding species to the British Isles; it nested at the site of some deliberate military inundations at Pett Level in Sussex (Cooke, 1946).

To sum up, birds live in a changing environment. Some of these changes are continuous, gradual or consistent trends in particular directions. Some are cyclical. Some are irregular. Some are irregular and cataclysmic. To all these changes the populations of birds respond, sometimes with a neat and astonishing rapidity. This leads us to suspect the existence of self-regulating mechanisms by which the population rapidly becomes adjusted to the new carrying capacity of the land. What these self-regulating mechanisms consist of we can scarcely begin to guess. But at least we can suggest that if there is one factor in the environment to which bird populations are more closely adjusted than any other, that factor is food.

Unfortunately it is beyond the scope of this book to describe the most important avian self-limiting mechanism, or at least what I believe to be so. These are the aggressive displays of the individuals of non-social species or the individual and social displays of flocks, particularly of breeding flocks. Elsewhere (Fisher, 1954) I have pursued the paradox that the holding of territories by birds may be a form of sociality. Territory and social aggression is the basis of the dispersal mechanism of birds, the method by which the limits of potential living space can be tried and tested freshly each year. There is a lot of evidence now that the pioneering of new areas (opened up, for instance, by a climatic amelioration) is carried out by young birds; and that the pioneering is done by individuals which are unable to stand the more dynamic display of the maturer birds, and thus tend to probe into new country, having less drive than their parents to return to the old home.

CHAPTER XI

BIRD NUMBERS AND MAN

THE hatred of gamekeepers and landowners for birds and beasts of prey is so strong, so deeply imbedded in their mental structure, that anyone who even questions the wickedness of corvids and hawks is usually branded by them as a mild form of lunatic. The notion that populations of several different species of predators can exist quite happily without depressing the numbers of the birds upon which they prey would arouse today good-humoured contempt and ridicule in the sporting papers; if resolutely maintained, the idea would send some gamekeepers and landowners I know into fits. Nevertheless, the evidence gradually accumulates that in the most stable natural communities, except in very special cases, and in most of the situations in which wild birds are preyed upon by wild birds, the numbers of predators have little, if any, effect upon the numbers of their prey. On the contrary, it is the numbers of the prey that affects the numbers of the predators.

One of the most recent exponents of this case is Errington (1946), though he is at pains to point out that this generalization is of the nature of a rule, which can be broken. He is also careful to point out that there are many special cases where predators do control, or have controlled, the numbers of their prey, even to the extent of destroying it and thus depriving themselves of all or much of their stake in the future. It is rash to generalize further, but I feel emboldened to suggest that the chief instances involving the breaking of this rule are when the predator is man, or one of the exotic animals which he takes with him on his way through the world, or introduces suddenly, and often unpredictably, into new communities. There are a few other predators which will kill the geese that lay the golden eggs—for instance, mammals belonging to the Canidae, particularly the wolves. But man has a record for exterminating his prey which is unparalleled in the animal kingdom, and the only other animals that come within any distance of him are a few enterprising animals parasitic upon the communities which he establishes, notably the rats. Yet it is not by preying directly upon them that man performs his most astonishing feats of animal control. The conscious and unconscious power which man can, and does, exercise, directly or indirectly, over the numbers of wild birds, can be analysed on present evidence in vast detail, and only at the end of such an analysis can the terrible power of man be believed.

A great deal of the work of E. M. Nicholson consists of an analysis of this relationship. It has culminated in his remarkable book *Birds and Men* (1951) and a stimulating paper (1952). Nicholson shows that man can affect bird populations by hunting, shooting, trapping and eating birds, by keeping them as pets, by watching them, by encouraging certain parasites and predators, by killing their food, removing their habitats, depriving them of water, changing the soil, altering water-levels or diverting rivers, by dumping his waste products or offal, by exposing his crops, by planting and gardening, by creating bird sanctuaries or preserving habitats.

It is an interesting fact that, at least in the North Atlantic area, in those places where man, as a hunter, still takes birds or eggs on any large scale for food, he has usually arrived at a stable relationship with his prey, at a situation of nice and prolonged balance. This state of affairs is described in some detail by Fisher and Lockley (1953), who show that sea-bird fowling now survives strongly only in the Faeroes, Iceland and Greenland, although till 1930 it flourished also at St. Kilda and is still rather desultorily carried on in the Outer Hebrides. The various fowling communities of these cliff-bound islands had often developed quite different methods of killing their prey, but all seemed to have arrived at a relationship which meant that they did not take more than half the annual output of young just before they were fledged. As far as can be detected from the available statistics of St. Kilda, the Faeroes and Iceland, there is no evidence at all that this cropping has any depressive effect upon the populations of the species concerned. In the case of the fulmars and puffins, where the young are taken just before fledging, the young are at this stage already deserted by their parents; this therefore inhibits their parents in no possible way from returning to the same site in any subsequent year. It is interesting to record that all these island wildfowling communities have very strict rules and lore about the right dates upon which to take a sea-bird, upon whether to take eggs[1] or young (they seldom take eggs of species like the fulmar, which do not lay replacements), and sometimes upon the rotation of the areas of bird cliff to be cropped. Each, as I have hinted, has arrived in his own way at a true biological relationship with his prey. It is interesting to observe that these primitives, the sea-fowlers, have done what the more 'advanced' sportsmen with their shotguns have not always done.

I shall leave those readers who wish to follow the way in which man has influenced animal populations by altering their habitat with Nicholson's analysis of the situation in Britain (1951) where, after all, there is scarcely a square mile but does not annually suffer the managing hand of man. Even in the remotest parts of the Scottish Highlands the landscape differs entirely from what it was before man took a hand. Upon man's agriculture and fisheries a growing hand of opportunist birds acts as scavengers and parasites. On land, birds haunt his midden and refuse tips, hang about his farmyards, follow his plough, raid his stacks, steal his crops, nip off the buds of his fruit-trees, drink, bathe and go fishing in his waters, roost in and foul his woods, nest on his houses and prey on his domestic animals. On land, the most harmful opportunist thief of human crops is unquestionably the wood-pigeon, which does very much more damage than the rook. The latter's position is doubtful, inasmuch as it can be proved to do a certain amount of harm by eating corn but may do a certain amount of good by eating insects, though it cannot be proved to do the latter. The house-sparrow is not so much a parasite of man as an associate of domestic beasts, particularly of horses. It is

[1]Recently Cott (1953–54) has published an analysis of the exploitation by man throughout the world of the eggs of wild birds. Man's worst mistakes have been in newly-opened countries, and "over-egging has contributed towards the extinction of some, and to the depletion to danger-point of several". But Cott clearly shows that national, planned egging can be, and is, operated without detriment to the population of certain species.

becoming gradually rare in London as internal combustion engines replace horses. Some birds exploit human waste-products. For instance, the existence of many a local ornithological society or bird section of a natural history society will probably be found to be correlated with the presence of a good local sewage farm; and Peterson (1950) points out that the best places to see rare gulls from Europe in New York is at the main sewage outfall at 92nd street in Brooklyn.

In 1936 Sæmundsson was the first, so far as I can discover, to suggest a general link between the extraordinary spread of the fulmar and the offal of man's whaling and trawling activities. We now know that the spread of the fulmar started in Iceland in the middle of the eighteenth century, more than 100 years before the start of trawling; so that the suggestion of Lockley and Marchant (1951) that the spread is simply due to the offal from modern trawling is an incomplete one. I have developed the connexion between the early years of the fulmar's spread and the pursuit of right whales in wooden ships in the Denmark Strait and the east Greenland waters (Fisher, 1952). But it was Sæmundsson who first thought of the connexion, and F. Faber (1824), who first suggested that the secret of the fulmar's spread would be found in the neighbourhood of Iceland, which he visited in 1819–21.

There are many other sea-birds besides fulmars which have become dependent on man's harvest of the sea. For a large part of the winter, many individuals of most species of British gulls gather at the fishing ports and become scavengers of the fish wharves and followers of the boats. For many, man, must be the prime provider. In the Antarctic, several of the local tube-nose species have become scavengers of the whaling yards and whaling ships, just as the fulmar was in the old whaling days in the north. In the case of the fulmar we know that the population has responded to the new supply of food.

When man decides to control the numbers of particular wild birds, he becomes like no ordinary predator; for he is skilful and ruthless enough to go on until sometimes he has killed the last bird or taken the last egg. At about the turn of the century, he had depressed to very small populations all but the most tenacious of Britain's birds of prey, and succeeded in extinguishing from our breeding list altogether the white-tailed eagle and the osprey. The last stages of the disappearance of these two birds were assisted by a particularly eager *côterie* of collectors, who were prepared to go on until there was nothing left to collect. Happily the days of the St. Johns, Booths and Colquhouns are nearly over, and while there are some very rare species with but a foothold in Britain, whose nesting-grounds are still in danger from and occasionally raided by collectors, the reverse process is now well under way. It is true that another generation or so of propaganda will be necessary before landowners and their keepers can learn to tolerate, and share with, all kinds of birds of prey. But the bird protection movement from the earliest days has always had the sympathy of at least the more enlightened landowners and sportsmen. Owing to the very existence of protection the societies owning and administering sanctuaries, conducting campaigns and circulating information, are able to point to species whose presence as British breeding birds depends on man the conservator. I have analysed the effects of sixty years of organized bird protection on the variety and numbers of the British avifauna (Fisher, 1951). During the nineteenth century this became impoverished, certainly in variety of species and probably also in numbers. It was to a great extent due to the invention of the shotgun, and to collectors, sportsmen, dealers and keepers, but a lot of it may have been due to the removal of habitats, by the draining of fens and marshes and the cutting of woods, and the spread of industry. At least nine species became totally extinct as nesters in Britain in the nineteenth century. Six of the nine were marsh-birds—Savi's warbler, bittern, black-tailed godwit, ruff, avocet and black tern. It is also possible, though not formally proved, that the melodious warbler bred in Britain in the nineteenth century. These species probably disappeared primarily because of the draining of fens, but not certainly for this reason. It is possible, for instance, that the avocets were finished off because their eggs were taken for human food. The other three non-marsh-birds that became extinct in the nineteenth century are the great auk, of which the last British example was probably killed in or about 1840, and the last of all killed in Iceland on about

4th June, 1844, the great bustard, which was shot, and the honey-buzzard, which was shot or robbed. Six of these species have nested again in the twentieth century, and two of them—the bittern and the avocet—have properly established themselves as the result of protection.

During the nineteenth century, to set against the losses, there were two introductions, the capercaillie (re-introduced) and the little owl, and the already introduced red-legged partridge was reinforced. There were also four colonizations, all ducks—gadwall, wigeon, pintail and goosander—of which at least the gadwall was helped by introduction. It is interesting to note that the gains in the nineteenth century were nearly all sporting birds and that they were outweighed by the losses.

Early in the present century two more species became extinct, as British nesters, the white-tailed eagle and the osprey. But the other birds of prey recovered from the depressed populations which consistent and enterprising human persecution had brought them to. In the present century the raven, carrion-crow, long-eared, brown and little owls have increased continuously, and within the last decade, particularly since the Second World War, ten other predators have started to increase. This may be partly due to the lack of keepers, for the magpie, which through keepering was extinct in Norfolk before the First World War, is now an abundant species there. The barn-owl has been recovering slowly. The peregrine, in spite of persecution in the Second World War to save carrier pigeons, has been spreading to sites not occupied for 100 years. The parkland falcon of the southern counties, the hobby, is building up its numbers after persecution, of which much has been by egg-collectors. The goshawk has recently bred, and may establish itself, in England. The merlin is increasing, the buzzard increasing rapidly in numbers and in area occupied; the golden eagle has chosen the middle of this century to extend its range for the first time for many years to the Lowlands of Scotland and Northern Ireland, and is increasing in numbers; and the hen-harrier is increasing in Orkney under close protection and is re-establishing itself in parts of the Scottish Highlands. Indeed, only three predators appear still to be decreasing.

The hooded crow's slow retreat into the Highlands and its replacement by the carrion-crow is probably linked with the ameliorating climate. Gamekeepers now concentrate their energies particularly on the sparrow-hawk, and few even of the most ardent protectionists object to this; and this woodland and hedgerow species, though still common, is probably slowly decreasing. The numbers of the marsh-harrier, *Circus aeruginosus*, are still ridiculously small and in some recent years there has been only one nest in Britain, and its numbers are even more precarious than those of the kite.

In the present century there has been so little change in the numbers of the breeding kites, *Milvus milvus*, in the British Isles (they are almost all in central Wales) that it is not possible to record an increase or a decrease. It has been on the brink of doom for over fifty years but has never fallen in. In the twentieth century, as far as I know, never more than ten pairs of kites have nested, or less than three; on an average about six. Yet in the last century the kite nested in every English county except one, in most of Wales and the Highlands, and in some of the counties in the Lowlands (Fisher, 1948).

It can be seen from this account of the birds of prey that, while the situation remains unsatisfactory with a handful of species, all interesting and beautiful, there has been a marked recovery in the present century. This can only be ascribed to the change in public opinion, and is living proof of the fact that the opinions which, half a century ago, were confined to a few eccentric ornithologists, who were no doubt then regarded as sentimental bird lovers, are now shared by the majority of people, even of those who look down the barrels of a shotgun. A ghost from the past, reading the correspondence columns of *The Times* in the 1950s, in which, almost every week, some question of nature protection is vehemently ventilated, would be as astonished as he would be gratified with the change in public opinion. Yet those who follow the fortunes of the Royal Society for the Protection of Birds, as they can by reading through a file of its journal, *Bird Notes*, can see that this change has been entirely gradual. My own idea of the success of the bird-protection societies in bringing about this change is that it was assured the moment bird-watching became regarded as an intellectual pursuit rather than an aimless entertainment. I would not be surprised if the pioneer books of the 'fact-finding' school, such as those of Nicholson (1927, 1931) did not, by starting us all bird-*watching*, do more for protection than the earlier, much more emotional and sometimes positively disingenuous propaganda of W. H. Hudson (1894).

But we must continue our profit-and-loss account of Britain's nesting birds. We have described the situation, on the whole encouraging, among the birds of prey and shown that what is good about it is probably due to protection. One of the most important factors in the protection of birds of prey has been the bounty system instituted by the Association of Bird Watchers and Wardens, and continued by the Royal Society for the Protection of Birds after their recent amalgamation. Under this system, rewards are offered to occupiers of land on which rare birds of prey can be proved to raise young to the free-flying stage. A very considerable sum of money is now annually expended on golden eagle and on kite bounties.

If we analyse the birds of woodland, agricultural land and built-up areas, we find that in the present century twice as many species have noticeably increased as have decreased. There is only one colonization to record, the black redstart, which invaded England early in the present century, possibly on a climatic spread, and is now a local breeder in built-up areas, particularly in ruins, in south-east England. The factors behind some of the decreases can be identified. That of the house-sparrow may follow the decrease of horses, that of the corn-crake the custom of earlier hay-mowing. None of the observed increases can be particularly attributed to protection, unless they owe something to the prohibition of selling cage birds caught wild in Britain: this may have helped the increase of every species of indigenous British finch, that has been observed.

If we follow the fate of birds of mountain, moorland and heath, we see that more than half of the species have noticeably decreased and only a few—grey wagtail, curlew and greenshank—have increased, Nobody yet seems to have thought of a satisfactory explanation of the interesting decline in the numbers of twite, ring-ousel, wheatear, stonechat, dunlin, golden plover and even red grouse. Among birds of marsh and fresh water we find half a dozen colonists in the present century, among waders the little ringed plover, and the avocet; the Slavonian and black-necked grebes, and of course the bittern, must be mentioned; the whooper swan has also nested. The only extinction in this group appears to be that of the Kentish plover, though there are rumours that it may still occupy one breeding place in south-east England. Some other species have noticeably decreased, such as the bearded tit, grey-lag goose, black-throated diver and spotted crake. The first and last of these are very sporadic in their numbers and movements. The other two need breeding grounds free from human disturbance, from which they increasingly suffer. Far more birds in this category have increased. Nearly every species of British breeding duck, for instance, shows an improvement in numbers, and this can be said to be due to protection, even though this is in many cases done for sport. Among the sea-birds, the noticeable increases have been among the gulls, many of which are parastic on the fishing industry, the fulmar, whose increase is analysed elsewhere (p. 166) and among a group which undoubtedly benefits from protection and which includes the eider, gannet and at least three species of tern, Sandwich, roseate and common. Fisher and Lockley (1953) show that Britain is now the European headquarters of the roseate tern (p. 131). The recovery of this species has been due to deliberate and careful local protection, and the location of most of the big colonies is carefully kept secret.

Some of the sea-birds which have decreased may have suffered from man. Why the Manx shearwater is apparently decreasing we do not know, but we suspect that the reduced numbers of the little tern may be due to the increase of bungalows along much of their natural nesting habitat, shingle banks above the high-tide mark. The noticeable decrease of all the British species of auk—razorbill, guillemot, puffin and probably also black guillemot—is not properly understood, but it is known that there is a high mortality from 'oiling' and that rats, which have established themselves on the islands where some breed, can and do severely depress auk populations.

To sum up on the question of protection. When the returned breeding avocet produces over 100 young ones a year within seven years of its first arrival, and does so exclusively (or almost exclusively) on a sanctuary belonging to the Royal Society for the Protection of Birds and watched day and night, we can begin to see that the Protection Society has no mere propaganda function. For the first time in history man is able, to a certain extent, by protection, administration and care, to direct and improve the composition of his native bird fauna. In the last fifty years in Britain, in spite of mistakes and muddling, our avifauna has regularly gained ground. Of course there are some changes which man seems powerless, or almost powerless, to influence, and some changes are due purely to nature. But I think it is mainly man's doing that we in Britain now have a greater population and variety of wild birds to enjoy than at any previous time of our history.

Man's power to look after and influence the numbers of wild birds is proved. Scarcely needing proof is his power to influence avifaunas by the simpler, more direct and sometimes dangerous method of introduction. A useful account of man's additions to the British fauna is that of Fitter (1950), who shows that about ten per cent of our present breeding land vertebrates—birds, mammals, reptiles and amphibians—are naturalized aliens. Fitter's analysis contains several grim warnings of the dangers of introduction, but these refer almost entirely to mammals. There is no evidence that any introduced bird has become a pest though some may feel doubtful still about the little owl. If we take a strict view of what constitutes a truly feral introduced species, we cannot embrace a very long list of birds, for we must ignore the many exotic species which breed free-winged in private avicultural collections or in collections, belonging to societies, such as Whipsnade Zoo or the Severn Wildfowl Trust. The mute swan and the pheasant probably represent ancient introductions. The French partridge probably established itself in the eighteenth century, though it was assisted by further introductions in the nineteenth. The capercaillie, once indigenous, became extinct and was successfully reintroduced at Taymouth in Perthshire in 1837; the story of its subsequent spread has been excellently told by Harvie-Brown (1888) and Pennie (1950–51). Pennie shows that this beautiful game-bird seems to have reached its peak in distribution and numbers at the end of the first decade of the present century, after which there has been very little advance and in some areas a decrease. Sportsmen and landowners have been under frequent temptations to introduce exotic game-birds, so much so that in their recent identification book, Fitter and Richardson (1952) have correctly figured many of these species because the amateur bird watcher may often encounter them. There are several estates where Amherst, golden, silver and Reeves's pheasant are at liberty, and there have been attempts to introduce the red jungle-fowl, the chukor (a relation of the French partridge) and even wild turkeys; some of these attempts have been temporarily successful. Exotic pigeons, too, have been kept free-winged, and there are several half-wild colonies of the little Barbary dove in south-east England which probably would not survive without feeding in winter but which otherwise look after themselves.

There have been a great many introductions of waterfowl. The Canada goose from North America has often been introduced to Britain from the seventeenth century onwards, and it is now breeding in so many counties of England, Scotland and Ireland that it is accepted to the British list as a wild species. As I mentioned elsewhere, other introductions, such as that of the gadwall, may have served to boost a small existing wild stock. In recent years the mandarin-duck, *Aix galericulata*, has established itself in wild self-supporting colonies, the most noticeable of which is perhaps that at Virginia Water in Surrey; and among the other anatids which are capable of supporting themselves at complete liberty, and occasionally do, are the Mandarin's close relation, the Carolina or wood-duck, the rosy-bill, the Egyptian and the Chinese goose, and the black swan.

Countries which are remoter than Britain is from continents with packed avifaunas, and which themselves start with impoverished or specialized avifaunas, have suffered far greater changes than has Britain as the result of introductions. New Zealand, for instance, has an altogether exceptional group of introduced species, the result not only of its original impoverished fauna, but also of having been settled, mostly during the last century, by colonists with the natural wish to surround themselves with some of the familiar animals of the lands from which they came. Between the years 1773, when Captain Cook introduced the fowl, and 1906, when the little owl arrived, no less than twenty-seven kinds of birds were successfully introduced into New Zealand. Many of these introductions took place between 1860 and 1870, and most between 1850 and 1906, when five different societies were formed with the specific purpose of introducing animals into the country (Thomson, 1922).

Introductions did not merely come from the original home of the colonists. Some of the pheasants came from China. Many birds came from Australia—for instance, the Australian magpie, the Australian quail and the black swan. The mynah came from India, the California quail from America; guinea-fowl, turkeys and peafowl were introduced from time to time and have gone wild on various parts of both islands. Their countries of origin are, respectively, Africa, America and India, though there is no doubt that the original birds in this case were domestic ones, probably from England. The English birds which were introduced, and which have successfully established themselves, include the chaffinch, blackbird, starling, sky-lark, song-thrush, hedge-sparrow, goldfinch, rook, house-sparrow, redpoll, bullfinch and siskin. I might also mention the European quail and the cirl-bunting.

Introductions of mammals into New Zealand were not always deliberate. As in most countries, other than those of their origin, the house-mouse and black rat found their way to New Zealand on ships, and many injurious insects were brought in the same way. But birds were introduced on purpose. Those colonists who desired to have near their homesteads the song of the lark little realized that this bird would become a pest in various parts of the island.

Even big continents with a heavy avifauna may offer unexploited niches to an introduced species of opportunist habits. There are places in North America inhabited by the introduced European goldfinch, European sky-lark and European tree-sparrow. In North America, however, the most spectacular spreads have been those of the introduced house-sparrow and starling. The English house-sparrow was taken to America and introduced into the neighbourhood of New York towards the end of the last century. It has now become of economic importance. There are excellent descriptions of the spread of the starling in Kalmbach (1938) and Bent (1950). Many attempts were made, beginning in 1872, deliberately to establish this bird, in Canada, Ohio, Massachusetts and New Jersey; but it was not until 1890 and 1891 that a starling introduction 'took', and even then, for six years, the bird was confined to the limits of greater New York. After that, to quote Bent,

> "the increasingly rapid spread began which, in little over a half century, took it to the Pacific coast. A study of the chronology of this spread shows that it was not a steady progression but somewhat spasmodic; a big expansion in one year followed by several years while the territory gained was filled in and little new occupied. How much transportation has helped in this spread is problematical, but some long jumps have suggested that birds may have been 'stowaways'."

The starling has been introduced into other parts of the world. It is thus established at Cape Town, South Africa, and is extending into the adjoining district (Roberts, 1940). Scott (1950) describes the introduction of starlings into Maracaibo, Venezuela, in 1949. Some starlings came on board a British ship in Falmouth on 29th October. The ship passed across the Atlantic through the Azores, and close to Martinique and Curaçoa; but the birds (five survivors of seven, the other two being eaten by the Chinese members of the crew) only left the ship when it was right inside the lagoon at Maracaibo on 12th November. On their voyage they lived on cockroaches, potatoes and fish.

It would be tedious and pointless to list the numerous cases of the introduction by man of bird species from one country to another. It is possible that some of the island birds that have been persistently referred to in the literature as 'obvious introductions' may really be of wild origin, since till recently the possibilities of drift introductions (see p. 105) have been widely overlooked. (In the same way, many birds that are obviously drift-borne wanderers have been recorded in the most unlikely places such as western Irish or Hebridean islands, as 'escapes from captivity'). Sometimes deliberate introduction has been done to preserve a species. Thus in 1911, when the plume trade was at its height, Sir William Ingram (1911) feared that the status of the greater bird of paradise in its home on Aru Island, New Guinea, was being seriously threatened by commercial interests. He accordingly bought forty-eight birds and established them on the island of Little Tobago in the West Indies, providing a watcher to look after them. They quickly adapted themselves to their new environment, and have bred and increased on its 400 acres. They have taken to a local fruit tree, the pawpaw, and seem to be in no danger from the local hawks. Guppy (1931) recorded that they were maintaining their numbers, but there is no mention of the birds in the paper of Belcher and Smooker (1934–37) on the birds of Trinidad and Tobago.

It can be seen from this chapter and its predecessors that man has fantastic power over wild birds, but that he is in no position to predict very accurately the consequences of the use of this power. We are only beginning to find out some of the factors that control and regulate bird population. Clearly, as the world contracts and man comes to dominate it even more, if he wishes to enjoy wild birds he must understand far more about their numbers. At present, he is not very good at counting birds, and even less good at deciding what to do when he has counted them. As Nicholson hints, we are only beginning to make a scientific study of our own administrative experience in trying to control wild bird populations. "Scientific bird protection," says Nicholson, "cannot become a practical possibility until much more knowledge of underlying problems has been gained". This, in my view, is unduly pessimistic, for, as we have already seen, scientific bird protection has been applied, albeit crudely, and is already paying dividends in Europe and North America. But anybody who reads Nicholson's stimulating paper or his book (1951) will be clear about what man is driving at. The bird protector of the future is going to do what he wants to do only if he learns something about ecology, and it will not help him or the birds very much if he puts on shining armour and leads an emotional crusade. The civilized world is no longer in need of emotional persuasion that birds are worth protecting.

BIBLIOGRAPHY

AHLMANN, H. W. (1949) *Geogr. J.* 112: 165–95.

ALEXANDER, C. J., and ALEXANDER, H. G. (1909) *Brit. Birds*, 2: 322–6.

ALEXANDER, W. B. (1932) *J. Anim. Ecol.* 1: 58–64. (1933) *Ibid.* 2: 24–35. (1945–47) *Ibis*, 87: 512–50; 88: 1–24, 159–79, 271–86, 427–44; 89: 1–28.

ALLEN, E. G. (1939) *Auk*, 56: 283–90. (1951) *Trans. Amer. Philos. Soc.* (n. ser.) 41: 385–591.

ALLEN, F. H. (1940) *Auk*, 57: 75–82.

AMADON, D. (1947) *Condor*, 49: 159–65. (1949) *Ibid.* 57: 250–8. (1950) *Bull. Amer. Mus. Nat. Hist.* 95: 157–262.

AMEGHINO, F. (1891) *Rev. argent. Hist. nat.* 1: 441–53.

ARCHEY, G. (1941) *Bull. Auckland. Inst. Mus.* no. 1; 145 pp.

AUDUBON, J. J. [L.] (1827–38) The birds of America. . . . London.

BAKER, R. H. (1951) *Univ. Kansas Publ. Mus. Nat. Hist.* 3: 359 pp.

BAUER, K. (1952) *J. Orn.* 93: 104–11.

BEEBE, C. W. (1918–22) A monograph of the pheasants. London, 4 vols. (vol. 2, 1921).

BEECHER, W. J. (1953) *Auk*, 70: 270–333.

BELCHER, C., and SMOOKER, G. D. (1934–37) *Ibis* (13) 4: 572–95; 5: 279–97; 6: 1–35, 792–813; (14) 1: 225–49, 504–50.

BELON DU MANS, P. (1554) *Les observations de plusieurs singularitez et choses memorables, trouvées en Grece, Asie, Judée, Egypte, Arabie et autres pays estranges.* Paris.

BENSON, C. W. (1942) *Bull. Brit. Orn. Cl.* 63: 8–19. (1946) *Ibis*, 88: 448–9.

BENT, A. C. (1923) *Bull. U.S. Nat. Mus. no.* 126; 244 pp. (1925) *Ibid.* no. 130; 316 pp. (1950) *Ibid.* no. 197; 411 pp. (pp. 182–214).

BEWICK, T., and BEILBY, R. (1797–1804) History of British birds. . . . Newcastle-upon-Tyne, 2 vols.

BLAKER, G. B. (1934) The barn owl in England and Wales. London.

BLANCHARD, C. E. (1859) *Ann. Sci. nat.* (4) 11: 11–145.

BODDAERT, P. (1783) *Table des Planches enluminéez d'Histoire naturelle.* Utrecht.

BOND, J. (1950–51) Check-list of birds of the West Indies. Philadelphia (200 pp. 1950; first suppl. 22 pp. 1951).

BRUCE, S. (1948) *Scot. Nat.* 60: 6–7.

BUICK, T. L. (1931) The mystery of the moa. . . . New Plymouth, New Zealand.

BURNS, F. L. (1901) *Wilson Bull.* 8: 84–103. (1914). *Bird Lore*, 17: 109–11.

BUXTON, A. (1940–44) *Trans. Norfolk Norw. Nat. Soc.* 15: 22–40, 150–9, 259–67, 332–41, 410–9.

CAIN, A. J. (1953) *Evolution*, 7: 76–83.

CAMPBELL, B. (1953) *Cardiff Nat. Soc. Rep. Trans.* 81: 4–65.

CAREW, R. (1602) The survey of Cornwall. London.

CATESBY, M. (1731–43) The natural history of Carolina, Florida, and the Bahama Islands: . . . London.

CHAPIN, J. P. (1936) *Rev. Zool. Bot. Afr. Tervueren*, 29: 1–6.

CHAPMAN, W. M. M. (1939) *J. Anim. Ecol.* 8: 286–99.

CHITTY, D. (1952) *Philos. Trans. Roy. Soc. London*, 236*B*: 505–52.
CLARKE, C. H. D. (1936) *Univ. Toronto Stud. Biol.* 41: 118 pp.
CLARKE, J. R. (1953) *J. Endocrin.* 9: 114–26.
CLAY, T. (1951) *Ibis*, 93: 628.
COCKRUM, E. L. (1952) *Wilson Bull.* 64: 140–59.
COMEAU, N. A. (1909) Life and sport on the north shore of the lower St. Lawrence and Gulf. . . . Quebec.
COOKE, R. (1946) *Brit. Birds*, 39: 71–2.
CORY, C. B., HELLMAYR, C. E., and CONOVER, B. (1918–48) *Field Mus. Publ.* (*Zool.*) 13: 11 numbers; 15 parts.
COTT, H. B. (1953–54) *Ibis*, 95: 409–49, 643–75; 96: 129–49.
COTTAM, C. (1939) *Tech. Bull. U.S. Dep. Agric.* no. 643: 140 pp.
CRIDDLE, N. (1930) *Canad. Field-Nat.* 44: 77–80.
CUMMING, W. A. (1953) *Notornis*, 5: 83–84.
DACKER, H. (1948) *Scot. Nat.* 60: 171–6.
DAMES, W. (1890) *Bih. Svensk. VetenskAkad. Handl.* 16: no. 4; 1.
DARWIN, C. (1859) The origin of species. . . . London.
DAUBENTON, E. L. (1765–81) [*Planches enluminees d'Histoire naturelle*]. Paris.
DELACOUR, J., and MAYR, E. (1945) *Wilson Bull.* 57: 3–55.
DELAMAIN, J. (1933) The days and nights of birds. London.
DERHAM, W. (1718) Philosophical letters between the late learned Mr. Ray and several of his ingenious correspondents, natives and foreigners. To which are added those of Francis Willughby, Esq.; . . . London.
DOMANIEWSKI, J. (1925) *Ann. Zool. Mus. Polon. Hist. Nat.* 4: 85–125.
DOUGLAS, N. (1928) Birds and beasts of the Greek Anthology. London.
DUFF, R. (1949) Pyramid Valley, Waikari, North Canterbury. Christchurch, N.Z.
DUNN, E. R. (1922) *Science*, 56: 336–8.
EBERHARD, J. (1948) *Oiseau*, 18: 117–32.
EDWARDS, G. (1743–51) A natural history of uncommon birds, . . . London. (1758–64) Gleanings of natural history, . . . London.
ELLIOTT, H. F. I. (1953) *Oryx*, 2: 41–53.
ELTON, C. (1927) Animal ecology. London. (1942) Voles, mice and lemmings. Oxford.
ERRINGTON, P. L. (1945) *Ecol. Monogr.* 15: 1–34. (1946) *Quart. Rev. Biol.* 21: 144–77, 221–45.
EVANS, A. H. (1903) Turner on birds: . . . Cambridge.
EVANS, D. (1948) *Irish Nat. J.* 9: 161–4.
EVANS, W. (1922) *Scot. Nat.* 1922: 9–12.
FABER, F. (1822) *Prodromus der isländischen Ornithologie.* Kobenhavn. (1824) *Isis*, 1824: 786.
FALLA, R. A. (1951) *Notornis*, 4: 97–100.
FISHER, J. (1939) Birds as animals. London. (1940) Watching birds. Harmondsworth. (1948) *Ann. Rep. Roy. Soc. Protect. Birds*, 1947: 22–34. (1951) *Trans. Norfolk Norw. Nat. Soc.* 17: 71–83. (1952) The fulmar. London. (1952b) *S. East Nat.* 57: 1–10. (1953) *Brit. Birds*, 46: 153–81. (1954) Evolution and bird sociality, pp. 71–83 of Huxley, Ford and Hardy (1954) *q.v.*
FISHER, J., and LOCKLEY, R. M. (1954) Sea-birds. London.
FISHER, J., and VEVERS, H. G. (1943–44) *J. Anim. Ecol.* 12: 173–213; 13: 49–62. (1951) *Proc. X. Int. Orn. Congr. Uppsala* 1950: 463–7.
FITTER, R. S. R. (1949) London's birds. London. (1950) *Discovery* 11: 58–62.

FITTER, R. S. R., and RICHARDSON, R. A. (1952) The pocket guide to British birds. London.

DE FLACOURT, E. (1661) *Histoire de la grande Isle Madagascar.* Paris.

FLEMING, C. A. (1939) *Emu,* 38: 380–413, 492–509.

FLEMING, C. A., and WODZICKI, K. A. (1952) *Notornis,* 5: 39–78.

FLEMING, C. A., and others. (1953) Check-list of New Zealand birds. Wellington.

FLEMING, G. (1871) Animal plagues. London, 2 vols.

FORBES, H. O. (1892) *Trans. Proc. N.Z. Inst.* 24: 185–9.

FORBES, S. A. (1907) *Bull. Ill. Nat. Hist. Surv.* 7: 305–35. (1913) *Ibid.* 9: 373–85.

FORBES, S. A., and GROSS, A. O. (1921–23) *Bull. Ill. Nat. Hist. Surv.* 14: 1–8, 187–218, 397–453.

FOSTER, R. J., BAXTER, R. L., and BALL, P. A. J. (1951) *Ibis,* 93: 53–9.

FOSTER-VESEY-FITZGERALD, D. (1953) *Oryx,* 2: 28–32.

FURBRINGER, M. (1888) *Untersuchungen zur Morphologie und Systematik der Vögel,* . . . Jena and Amsterdam, 2 vols.

GADOW, H. F. (1892) *Proc. Zool. Soc. Lond.,* 1892: 229–56. (1893) *Vögel, in* H. G. Brönn, *Die Klassen und Ordnungen des Thier-Reichs.* . . . Leipzig and Heidelberg, 2 vols. (1896) *Ibis,* (7) 11: 586–7.

GAUSE, G. F. (1934) The struggle for existence. Baltimore.

GEOFFROY SAINTE-HILAIRE, Etienne. (1807) *Ann. Mus. Hist. nat. Paris,* 10: 342–65. (1824) *Système dentaire des Mammifères et des Oiseaux.* Paris. (1833) *Nouv. Ann. Mus. Hist. nat. Paris,* (3) 2: 1–22.

GEOFFROY SAINTE-HILAIRE, Isidore. (1851) *Ann. Sci. nat.* (3) 14: 206–16.

GIBB, J., and GIBB, C. (1951) *Brit. Birds,* 44: 158–63.

GIBSON-HILL, C. A. (1948) *J. Soc. Pres. Faun. Emp.* (n. ser.) no. 58; 27–9.

GLADSTONE, H. S. (1910) The birds of Dumfriesshire, . . . London. (1911) Addenda and corrigenda to *idem.* Dumfries. (1922) Notes on the birds of Dumfriesshire. Dumfries.

GLEGG, W. E., and others. (1952) Check-list of the birds of Great Britain and Ireland. London.

GLEIZE, D. A. (1945) *Bull. Mass. Audubon Soc.* 29: 220.

GLOGER, C. W. L. (1834). *Vollständiges Handbuch der Naturgeschichte der Vögel Europa's.* Breslau, 600 pp.

GMELIN, J. F. (1788–93) *Systema Naturae.* Leipzig, 13th ed.

GROSS, A. O. (1944) *Auk,* 61: 513–37. (1945) *Auk,* 62: 241–56.

GUDHMUNDSSON, Finnur (1951) *Proc. X. Int. Orn. Congr. Uppsala* 1950 502–14.

GUPPY, P. L. (1931) *Avicult. Mag.* (4) 9: 5–8.

GURNEY, J. H. (1913) The gannet. . . . London. (1921) Early annals of ornithology. London.

VON HAAST, J. (1886) *Trans. Zool. Soc.* 12: 161–9.

HACHISUKA, M. (1937) *Tori,* 9: 461–3.

HACHISUKA, M., and DELACOUR, J. (1937) *Bull. Brit. Orn. Cl.* 57: 66–7.

HALLER, K. W. (1940) *Cardinal,* 5: 49–52, 1 pl.

HANSON, H. C., and GRIFFITH, R. E. (1952) *Bird-Banding,* 23: 1–22.

HANSON, H. C., and SMITH, R. H. (1950) *Bull. Illinois Nat. Hist. Surv.* 25: 59–210.

HANTZSCH, B. (1905) *Beitrag zur Kenntnis der Vogelwelt Islands.* Berlin.

HARRISON, W. (1577) Translation of *The Description of Scotland,* by

Hector Boece; and *The Description of England*: in *Chronicles of England, Scotlande and Irelande*, by Raphaell Holinshed. London.

HARRISSON, T. H. (1931) *London Nat.* 1930: 82–120.

HARRISSON, T. H., and HOLLOM, P. A. D. (1932–33) *Brit. Birds*, 26: 62–92, 102–31, 142–55, 174–95, 289–91.

HARRISSON, T. H., and HURRELL, H. G. (1933) *Proc. Zool. Soc. London* (A) 103: 191–209.

HARTERT, E., and STEINBACHER, F. (1903–38). *Die Vögel der paläarktischen Fauna.* . . . Berlin, 3 vols. and 8 suppls.

HARTING, J. E. (1866) The birds of Middlesex. . . . London.

HARVIE-BROWN, J. A. (1888) The capercaillie in Scotland. Edinburgh.

HAVERSCHMIDT, F. (1949) The life of the white stork. Leiden.

HEILMAN, G. (1926) The origin of birds. London.

L'HERMINIER, F. J. (1827) *Mem. Soc. linn. Paris*, 6: 1–93. (1828) Revised and enlarged ed., *v.* A. Newton (1884, p. 21).

HESSE, R. (1922) *S. B. naturh. Ver. preuss. Rheinl. Westf.* 1922: 4 pp.

HOLLOM, P. A. D. (1940) *Brit. Birds*, 33: 202–21, 230–44; 34: 93.

HOLMES, P. F., and KEITH, D. B. (1936) *Ibis*, (13) 6: 322–30.

HOPKINS, G. H. E. (1942) *Ibis*, (14) 6: 94–106.

HØRRING, R., SCHEEL, H., and TANING, Å. V. (1925–31) *Danmarks Fugle*, . . . København, 3 vols.

HOWARD, H. (1947) *Condor*, 49: 10–13. (1950). *Ibis*, 92: 1–21.

HOWELL, A. H. (1919) *Auk*, 36: 86–7.

HUDSON, W. H. (1894) *Roy. Soc. Protect. Birds Pamph.* no. 14; 32 pp.

HUNTINGTON, E. (1940) Principles of economic geography. New York. (1945) Mainsprings of civilization. New York and London.

HUTCHINSON, G. E. (1950) *Bull. Amer. Mus. Nat. Hist.* 96: 554 pp.

HUTTON, F. W. (1892) *Trans. Proc. N.Z. Inst.* 24: 93–172.

HUXLEY, J. S. (1939) *Bijdr. Dierk.* 27: 491–520. (1942) Evolution the modern synthesis. London.

HUXLEY, J. S., FORD, E. B., and HARDY, A. C. (ed.) (1954) Evolution as a process. London.

HUXLEY, T. H. (1867) *Proc. Zool. Soc. Lond.* 1867: 415–72; and *Sci. Mem.* 3: 238–97. (1868) *Proc. Zool. Soc. Lond.* 1868: 294–319.

VAN IJZENDOORN, A. L. J. (1950) The breeding-birds of the Netherlands. Leiden.

INGRAM, W. (1911) *Avicult. Mag.* (3) 2: 142–7.

JESPERSEN, P. (1949) *Rapp. Cons. Explor. Mer.* 125: 36–9.

JOHANSEN, H. (1946) *Dansk Orn. Foren. Tidsskr.* 40: 121–42.

JOHNSEN, S. (1929) *Bergens Mus. Årb.* no. 2; 5–118.

JOHNSON, R. A. (1940) *Bird-banding*, 11: 1–17.

JOLLIE, M. (1953) *Ibis*, 95: 369–71.

JONES, L. (1898) *Wilson Orn. Chapter Gen. Notes Bull.* 18: 5–9.

JONSTON, J. (1650) *Historia Naturalis de Avibus*. Frankfurt-am-Main.

JOURDAIN, F. C. R. and WITHERBY, H. F. (1918) *Brit. Birds*, 11: 266–71; 12: 26–35.

KALELA, O. (1946) *Orn. Fenn.*, 23: 77–98. (1946b) *Ann. Acad. Sci. Fenn.* (A) 4: no. 12; 81 pp. (1949) *Bird-Banding*, 20: 77–103. (1950) *Orn. Fenn.* 27: 1–30. (1952) *Fennia*, 75: 38–51.

KALMBACH, E. R. (1928) *Dep. Bull. U.S. Dep. Agric.* no. 1571; 26 pp.

KENDEIGH, S. C. (1944) *Ecol. Monogr.* 14: 67–106.

KEVE, A. (1948) *Ibis*, 90: 465–6.

KEVE, A., and UDVARDY, M. D. F. (1951) *Proc. X. Int. Orn. Congr. Uppsala* 1950: 468–76.

KING, R. T. (1937) *J. Forestry*, 35: 523–32.
KLEINSCHMIDT, O. (1941) *J. Orn.* 89: 157–8.
KLOSTER, R. (1921) *Norsk Jaeg.-Fisk.-Foren. Tidsskr.* 50: 317–32.
KOFORD, C. B. (1953) *Nat. Audubon. Soc. Research Rep.* no. 4; 154 pp.
KORTRIGHT, F. H. (1943) The ducks, geese and swans of North America. Washington.
LACK, D. (1935) *J. Anim. Ecol.* 4: 43–51. (1937) *Ibis*, (14) 1: 369–95. (1945) *J. Anim. Ecol.* 14: 12–16. (1946) *Brit. Birds*, 39: 204–6. (1947) Darwin's finches. Cambridge. (1947b) *Ibis*, 89: 302–52. (1950) Robin redbreast. Oxford.
LACK, D., and ROBERTS, B. B. (1934) *Ibis*, (13) 4: 799–807.
LACK, D., and VENABLES, L. S. V. (1939) *J. Anim. Ecol.* 8: 39–71.
LAMBRECHT, K. (1933) *Handbuch der Palaeornithologie*. Berlin.
LANHAM, U. N. (1947) *Auk*, 64: 65–70.
LATHAM, J. (1781–85, suppls. 1787, 1801) A general synopsis of birds. London. (1790, suppl. 1801) *Index ornithologicus, sive systema ornithologiae*. London. (1821–28). A general history of birds. Winchester.
DE LAVALLE Y GARCIA, J. A. (1917) *Mem. Compañia Admin. Guano, Lima*, 8: 61–88.
LEIGH, C. (1700) The natural history of Lancashire, Cheshire, and the Peak in Derbyshire, . . . Oxford and London.
LEOPOLD, A., and BALL, J. N. (1931) *Canad. Field-Nat.* 45: 162–7.
LEOPOLD, A. S., and SMITH, R. H. (1953) *Calif. Fish & Game*, 39: 95–101.
LEWIS, H. F. (1929). The natural history of the double-crested cormorant (*Phalacrocorax auritus auritus* (Lesson)). Ottawa. (1941) *Auk*, 58: 360–3. (1941b) *Wilson Bull.* 53: 22–30.
LINCOLN, F. C. (1930) *Circ. U.S. Dep. Agric.* no. 118; 4 pp.
LINNAEUS, C. (1758) *Systema Naturae*. Holmiae, 10th ed.
LOCKLEY, R. M. (1952) *Ibis*, 94: 144–57.
LOCKLEY, R. M., and MARCHANT, S. (1951) *Brit. Birds*, 44: 373–83.
LØPPENTHIN, B. (1932) *Vidensk. Medd. naturh. Foren. Kjob.* 94: 147–86. (1948) *Ibis*, 90: 86–90.
LOTKA, A. J. (1934) *Act. Sci. Industr.* no. 187; 45 pp.
LOW, G. C. (1929) List of the vertebrated animals exhibited in the gardens of the Zoological Society of London, 1828–1927. Vol. II. Birds. London.
LOWE, P. R. (1949) *Ibis*, 91: 102–4.
LUMSDEN, W. H. R., and HADDOW, A. J. (1946) *J. Anim. Ecol.* 15: 35–42.
LUNDEVALL, C.-F. (1953) *Fair Isle Bird Obs. Bull.* no. 9; 5–12, no. 10; 11–18.
LYDDEKER, R. (1891) Catalogue of the fossil birds in the British Museum. London.
MACKENZIE, J. M. D. (1952) *J. Anim. Ecol.* 21: 128–53.
MÄCHLER, G. (1952) *Orn. Beob.* 49: 134–5.
MANLEY, G. (1952) Climate and the British scene. London.
MARCHANT, S. (1952) *Brit. Birds*, 45: 22–7.
MARSH, O. C. (1870) *Amer. J. Sci.* (2) 49: 205–17. (1872) *Ibid.* (3) 3: 56–7. (1872b) *Ibid.* (3) 3: 360–5. (1872c) *Ibid.* (3) 4: 344. (1872d) *Ibid.* (3) 5: 74. (1873) On a new sub-class of fossil birds (*Odontornithes*). New Haven. (1880) *Odontornithes*: a monograph on the extinct toothed birds of North America. Washington. (1893) *Amer. J. Sci.* (3) 45: 81–2.
MARTIN, M. (1698) A late voyage to St. Kilda. London.
MATHESON, C. (1953) *Brit. Birds*, 46: 57–64.

MATHEWS, G. M. (1927–31) *Systema avium Australasianarum*. London.

MAY, R., and FISHER, J. (1953) *Brit. Birds*, 46: 51–5.

MAYFIELD, D. H. (1953) *Auk*, 70: 17–20.

MAYR, E. (1926) *J. Orn.* 74: 571–671. (1940) *Amer. Nat.* 74: 249–78. (1942) Systematics and the origin of species from the viewpoint of a zoologist. New York. (1946) *Auk.* 63: 64–9. (1946b) *Wilson Bull.* 58: 1–41. (1951). *Proc. X. Int. Orn. Congr. Uppsala* 1950: 91–131.

MAYR, E., and AMADON, D. (1951) *Amer. Mus. Novit*, no. 1496; 42 pp.

MAYR, E., and VAURIE, C. (1948) *Evolution*, 2: 238–65.

MEINERTZHAGEN, R. (1930) Nicoll's birds of Egypt. London. (1952) *Ibis*, 94: 686–7.

MEISE, W. (1928) *J. Orn.* 76: 1–203. (1938) *Proc. VIII Int. Orn. Congr. Oxford* 1934: 49–189. (1938b) *Proc. IX. Congr. Orn. Int. Rouen*, 1938: 46–51.

MERREM, B. (1816) *Abh. preuss. Akad. Wiss.* 1812–13: 237–59. (1819) *Ibid.* 1819: 179–98.

MERRET, C. (1667) *Pinax rerum naturalium*, . . . London. (first ed. 1666).

VON MEYER, H. (1861) *Jahrb. Mineral.* 1861: 561.

MIDDLETON, A. D. (1934) *J .Anim. Ecol.* 3: 231–49.

MILLER, L., and HOWARD, H. (1948) *Publ. Carneg. Instn.* 584: 201–28.

MILNE-EDWARDS, A., and GRANDIDIER, A. (1866–73) *Recherches sur la faune ornithologique éteinte des Iles mascareignes et de Madagascar.* Paris, pp. 85–114.

MOLTONI, E. (1938) *Orn. Monatsb*, 46: 80–3.

MONTAGU, G. (1802) Ornithological dictionary; . . . London. (1813) Supplement to *idem.* London.

MORAN, P. A. P. (1952) *J. Anim. Ecol.* 21: 154–8.

MORE, A. G. (1865) *Ibis* (n. ser.) 1: 1–27, 119–42, 425–58.

MOREAU, R. E. (1930) The birds of ancient Egypt, vol. 1: 58–77 of R. Meinertzhagen (1930) *q.v.* (1952) *J. Anim. Ecol.* 21: 250-71.

MORENO, F. P., and MERCERAT, A. (1891) *An. Mus. La Plata Sec. Paleont.* part 1.

MORLEY, A. (1940) *J. Anim. Ecol.* 9: 84–8.

MORTON, J. (1712) The natural history of Northamptonshire; . . . London.

MOSBY, H. S., and HANDLEY, C. O. (1943) The wild turkey in Virginia : . . . Richmond.

MUIRHEAD, G. (1889–95) The birds of Berwickshire. . . . Edinburgh, 2 vols., vol. 1: 216–39.

MULLER, J. P. (1845–46) *Abh. preuss. Akad. Wiss.* 1845: 321–92, 405–6; reprinted (1847) Berlin, 71 pp.

MURPHY, R. C. (1936) Oceanic birds of South America. New York, 2 vols. (1938) *Science N.Y.* (n. ser.) 88: 533–9. (1949) A new species of petrel from the Pacific, pp. 89–91 of *Ornithologie als biologische Wissenschaft.* Heidelberg.

MURPHY, R. C., and MOWBRAY, L. S. (1951) *Auk*, 68: 266–80.

NASH, J. K. (1928) *Scot. Nat.* 1928: 69–75.

NEWTON, A. (1884) Ornithology *in Encyclopaedia Britannica*. London, 9th ed., vol. 18: 2–50.

NEWTON, A., GADOW, H. and others. (1896) A dictionary of birds. London.

NICHOLSON, E. M. (1927). How birds live. London. (1929) The study of birds. London. (1929b) *Brit. Birds*, 22: 270–323, 334–72. (1931) The art of bird watching. London, pp. 91–131. (1932) *Daily Mail*, 29 September. (1951) Birds and men. London. (1952) *Bull. Brit. Orn. Cl.* 72: 28–34.

NICHOLSON, E. M., and NICHOLSON, B. D. (1930) *J. Ecol.* 18: 51–66.

NITZSCH, C. L. (1811) *Osteografische Beiträge zur Naturgeschichte der Vögel.* Leipzig. (1829) *Observationes de Avium Arteria carotide communi.* Halle. (1833) *Pterylographiae Avium Pars prior.* Halle.

NITZSCH, C. L., and BURMEISTER, C. C. H. (1840) *System der Pterylographie. . . .* Halle. (1867) Pterylography. London, transl. of (1840) by W. S. Dallas, ed. P. L. Sclater.

NOPCSA, F. (1907) *Proc. Zool. Soc. Lond.* 1907: 223–36.

NORDENSKIÖLD, N. E. (1929). The history of biology. A survey. London, transl. (from Swedish) by L. B. Eyre.

OWEN, R. (1839) *Proc. Zool. Soc. Lond.* 1839: 169–71. (1842) *Trans. Zool. Soc. Lond.* 3: 29–32. (1843) *Proc. Zool. Soc. Lond.* 1843: 8–10. (1863) *Philos. Trans. Roy. Soc.* 153: 33–47. (1873) *Quart. J. Geol. Soc.* 29: 511–22. (1878) *Ibid.* 34: 124–30.

PALMER, R. S. (1949) *Bull. Mus. Comp. Zool.* 102: 656 pp. (1950) *Bull. Maine Audubon Soc.* 6: 20 pp.

PARKER, W. K. (1860) *Trans. Zool. Soc. Lond.* 2: 269–351.

PEITZMEIER, J. (1951) *Proc. X. Int. Orn. Congr. Uppsala,* 1950: 477–83.

[PENNANT, T.] (1766) The British zoology. London. (1784–85) Arctic zoology. London, 2 vols.

PENNIE, I. D. (1950–51) *Scot. Nat.* 62: 65–87, 157–78; 63: 4–17, 135.

PETERS, H. S., and BURLEIGH, T. D. (1951) The birds of Newfoundland. St. John's, 431 pp.

PETERS, J. L. (1931–51) Check-list of birds of the world. Cambridge, Mass., 7 vols. by 1951.

PETERSON, R. T. (1950) Birds over America. New York.

PHILBY, H. St.J. B. (1937) *Bull. Brit. Orn. Cl.* 58: 29–31.

PLOT, R. (1678) The natural history of Oxfordshire, . . . Oxford. (1686) The natural history of Staffordshire, . . . Oxford.

POULTON, E. B. (1903) *Proc. Ent. Soc. London* 1903: 1xxvii–cxvi.

PRESTON, A. W. (1891) *Trans. Norfolk Norw. Nat. Soc.* 5: 191–6.

PYCRAFT, W. P. (1910) A history of birds. London, 458 pp.

RAND, A. L. (1936) *Amer. Mus. Novit.* no. 890; 14 pp.

RAND, A. L., and TRAYLOR, M. A. (1950) *Auk,* 67: 169–83.

RAVEN, C. E. (1942) John Ray. Cambridge. (1947) English naturalists from Neckam to Ray. . . . Cambridge.

RAY, J. (1678) The ornithology of F. Willughby, . . . London.

REGAN, C. T. (1926) *Rep. Brit. Ass.* 1925: 75.

REICHENBACH, H. G. L. (1849–50) *Avium systema naturale. . . .* Dresden and Leipzig.

RENSCH, B. (1929) *Das Prinzip geographischer Rassenkreise und das Problem der Artbildung.* Berlin. (1931) *Proc. VII. Int. Orn. Congr. Amsterdam,* 1930: 197–205. (1933) *Verh. dtsch. Zool. Ges.* 1933: 19–83. (1936) *Arch. Naturges.* (n. ser.) 5: 317–63. (1938) *Proc. VIII. Int. Orn. Congr. Oxford,* 1934: 285–311. (1941) *Orn. Montasb.* 49: 94–104.

ROBERTS, A. (1940) The birds of South Africa. London and Johannesburg.

ROTHSCHILD, M., and CLAY, T. (1952) Fleas, flukes and cuckoos a study of bird parasites. London.

ROTHSCHILD, W. (1907) Extinct birds. . . . London.

ROWAN, M. K. (1952) *Ibis,* 94: 97–121.

RZACZYNSKI (1721) *Historia naturalis curiosa regni Poloniae.* Sandomirsk.

SÆMUNDSSON, B. (1936) *Islensk Dýr III. Fuglarnir* (*Aves Islandiae*). Reykjavík.

SAINTY, J. E., MOSBY, J. E. G., BUXTON, A., and ELLIS, E. A. (1939) *Trans. Norf. Norw. Nat. Soc.* 14: 334–90.

SALOMONSEN, F. (1944) *Göteborgs Vetensk. Samh. Handl. 3B*: 1–138. (1948) *Dansk Orn. Foren. Tidsskr.* 42: 85–99. (1950–51) *Gronlands Fugle. Kobenhavn.* (1951b) *Proc. X. Int. Orn. Congr. Uppsala*, 1950: 515–26.

SCHENK, J. (1926) *J. Orn.* 74: 106–21. (1944) *Aquila*, 50: 260–4.

SCHIØLER, E. L. (1931) Vol. 3: 357–83 of Hørring, Scheel and Tåning (1925–31), *q.v.*

SCHORGER, A. W. (1944) *Trans. Wis. Acad. Sci. Arts Lett.* 35: 1–59.

SCHULZ, H. (1947) *Die Welt der Seevögel ein Führer durch die Vogelbrutstätten der deutschen Küsten.* Hamburg.

SCHUZ, E. (1936) *Orn. Monatsb.* 44: 33–41.

SCHWENCKFELD, C. (1603) *Aviarium Silesiæ*, vol. 4 of *Theristopheum Silesiæ. . . .* Leipzig.

SCLATER, P. L. (1858) *J. Linn. Soc.* (*Zool.*) 2: 130–45. (1880) *Ibis*, (4) 4: 340–50, 399–411.

SCLATER, P. L., WHARTON, H. T., and four others. (1883) A list of British Birds. . . . London.

SCLATER, W. L. (1924–30) *Systema avium Aethiopicarum*; . . . London.

SCOPOLI, G. A. (1786–88) *Deliciae florae et faunae insubricae*, . . . Ticini.

SCOTT, P. (1949) *Severn Wildf. Trust Ann. Rep.* 2: 48 pp. suppl. (1950) *Brit. Birds*, 43: 369. (1950–51). *Severn Wildf. Trust Ann. Rep.* 3: 121–30; 4: 109–18. (1951). *Proc. X. Int. Orn. Congr. Uppsala*, 1950: 611–12. (1951b) *Ibid.*: 648–51. (1953) *Severn Wildf. Trust Ann. Rep.* 5: 64–7.

SCOTT, P., and FISHER, J. (1953) A thousand geese. London.

SCOTT, P., FISHER, J., and GUDHMUNSSON, F. (1953) *Severn Wildf. Trust Ann. Rep.* 5: 78–115.

SEELEY, H. G. (1866) *Ann. Mag. Nat. Hist.* (3) 18: 109–10. (1881) *Geol. Mag.* (n.s.) 8: 454–5.

SHARPE, R. B., and others. (1874–98) Catalogue of the birds in the British Museum. Lond, 27 vols. (1899–1909) Hand-list of the genera and species of birds in the British Museum. London, 5 vols.

SHUFELDT, R. W. (1915) *New Haven Conn. Trans. Acad. Arts Sci.* 19: 110 pp. (p. 41).

SIBBALD, R. (1684) *Scotia illustrata sive Prodromus historiae naturalis.* . . . Edinburgh, 3 vols.

SIMPSON, G. G. (1946) *Bull. Amer. Mus. Nat. Hist.* 87: 1–99.

SITWELL, S., BUCHANAN, H., and FISHER, J. (1953) Fine bird books 1700–1900. London and New York.

SMITH, S. (1950) The yellow wagtail. London.

SNYDER, L. L. (1935) *Univ. Toronto Stud. Biol.* no. 40; 66 pp.

SOPER, J. D. (1942) *Proc. Boston Soc. Nat. Hist.* 42: 121–5.

SOUTHERN, H. N. (1945) *J. Genet. London*, 47: 51–7. (1951) *Proc. Zool. Soc. Lond.* 121: 657–71.

SOUTHERN, H. N., and REEVE, E. C. R. (1942) *Proc. Zool. Soc. Lond.* (A) 111: 255–76.

STEERS, J. A. (1953) *Geogr. J.* 119: 281–98.

STEGMANN, B. (1934) *J. Orn.* 82: 340–80.

STEJNEGER, L. H. (1884) *Sci. Rec.* 2: 154–5. (1885) Birds, *in* The Standard Natural History. Boston, ed. J. S. Kingsley, vol. 4: 558 pp.

STEVEN, G. A. (1933) *J. Marine Biol. Assoc.* 19: 277–92.

STEWART, W. (1923) *Scot. Nat.* 1923: 141–6. (1930) *Ibid.* 1930: 15–21.

STOLPE, M. (1935) *J. Orn.* 83: 115–28.

STONE, W., and others. (1931) Check-list of North American birds. . . . Lancaster, Pa.

STONOR, C. R. (1939) *Bull. Brit. Orn. Cl.* 59: 57–61.

STRESEMANN, E. (1926) *Orn. Monatsb.* 34: 59–61. (1927–34) *Aves, in* Kükenthal, W. and Krumbach, T. *Handbuch der Zoologie.* Berlin and Leipzig, vol. 7, second half, 899 pp. (1939) *J. Orn.* 87: 312–425. (1948) *Orn. Ber.* 4: 193–222. (1951) *Der Entwicklung der Ornithologie von Aristoteles bis zur gegenwart.* Berlin.

STRESEMANN, E., and TIMOFÉEFF-RESSOVSKY, N. W. (1947) *Biol. Zbl. Leipzig,* 66: 57–76.

STRICKLAND, H. E. (1849) *Ann. Mag. Nat. Hist.* (2) 4: 338–9.

STRONG, R. M. (1939–46) *Field Mus. Publ.* (Zool.) 25: parts 1, 2, 937 pp.; 3, 528 pp.

SUNDEVALL, C. J. (1872–73) *Methodi Naturalis Avium Disponendarum Tentamen.* Stockholm; and revisions (1874) *K. Svensk Vetensk.Akad. Förh.* no. 2; 21–30: no. 3; 27–30.

SUTTON, G. M. (1942) *Cardinal,* 5: 151–4.

THOMPSON, B. H. (1932) *Contr. U.S. Wildlife Div. Occ. Pap.* no. 1; 82 pp.

THOMPSON, D'A. W. (1936) A glossary of Greek birds. London.

THOMSON, G. M. (1922) The naturalisation of animals and plants in New Zealand. Cambridge.

TICEHURST, C. B. (1938) A systematic review of the genus *Phylloscopus.* London.

TICEHURST, N. F., and HARTLEY, P. H. T. (1948) *Brit. Birds,* 41: 322–34.

TICEHURST, N. F., WITHERBY, H. F., and HAWKE, E. L. (1940) *Brit. Birds,* 34: 118–32, 142–55.

TIMOFÉEFF-RESSOVSKY, N. W. (1940) *J. Orn.* 88: 334–40.

TRISTRAM, H. B. (1911) The natural history of the Bible. London, 10th ed.

VAN TYNE, J. (1951) *Proc. X. Int. Orn. Congr. Uppsala,* 1950: 537–44.

UDVARDY, M. D. F. (1951) *Oikos,* 3: 98–123.

VÄLIKANGAS, I. (1951) *Proc. X. Int. Orn. Congr. Uppsala,* 1950: 527–31.

VERNER, W. (1914) *Country Life,* 36: 114–18.

VOGT, W. (1940) *Bol. Compañia Administradora Guano Lima,* 16: no. 10; 307–29. (1941) *Ibid.* 17: 127–68. (1942) *Ibid.* 18: no. 3; 3–132.

VOLTERRA, V., and D'ANCONA, U. (1935) *Act. Sci. Industr.* no. 243; 97 pp.

VOOUS, K. H. (1949) *Condor,* 51: 52–81. (1940b) *Ardea,* 37: 113–22. (1950) *Limosa,* 23: 281–92. (1953) *Beaufortia,* no. 30; 41 pp.

VOOUS, K. H., and VAN MARLE, J. G. (1953) *Ardea,* 41: extra number; 68 pp.

WAGNER, A. (1861) *S. B. bayer. Akad. Wiss.* 2: 146–54. (1862) *Ann. Mag. Nat. Hist.* (3) 9: 261–7.

WALLACE, A. R. (1859) *Ibis,* 1: 449–54. (1864) *Ibis,* 6: 36–41. (1876) The geographical distribution of animals. . . . London, 2 vols.

WETMORE, A. (1934) *Smithson. Misc. Coll.* 89: 1–11. (1951) *Ibid.* 117: no. 4; 22 pp. (1951b) *Proc. X. Int. Orn. Congr. Uppsala,* 1950: 51–74.

WHITE, G. (1947) The natural history of Selborne. London, ed. J. Fisher (first ed. 1789).

WILLIAMS, G. R. (1953) *Ibis,* 95: 676–92.

WILLIAMSON, K. (1948) The Atlantic Islands a study of the Faeroe life and scene. London.

WILSON, A., and ORD, G. (1808–14) American ornithology; . . . Philadelphia, 9 vols.

WITHERBY, H. F. (1941) Check-list of British birds. London.

WITHERBY, H. F., and JOURDAIN, F. C. R. (1929) *Brit. Birds*, 23: 154–8.

WITHERBY, H. F., JOURDAIN, F. C. R., TICEHURST, N. F., and TUCKER, B. W. (1938–41) The handbook of British birds. London, 5 vols.

WOOD, C. A. (1931) An introduction to the literature of vertebrate zoology. . . . London (ed.). (1942) *De arte venandi cum avibus* of Frederick II. Stanford University Press.

WOOLLEY, L. (1942) Ur of the Chaldees. . . . Harmondsworth, Penguin Books (first ed. 1929).

WYNNE-EDWARDS, V. C., LOCKLEY, R. M., and SALMON, H. M. (1936) *Brit. Birds*, 29: 262–76.

ZIMMER, J. T. (1926) *Field Mus. Publ.* (Zool.) 16: 706 pp.

ZIMMER, J. T., and MAYR, E. (1943) *Auk*, 60: 249–62.

INDEX

For Product Safety Concerns and Information please contact our EU
representative GPSR@taylorandfrancis.com
Taylor & Francis Verlag GmbH, Kaufingerstraße 24, 80331 München, Germany

www.ingramcontent.com/pod-product-compliance
Ingram Content Group UK Ltd.
Pitfield, Milton Keynes, MK11 3LW, UK
UKHW021501240926
475702UK00035B/58

* 9 7 8 1 0 4 1 3 5 2 9 3 8 *